MOON

The Earth series traces the historical significance and cultural history of natural phenomena. Written by experts who are passionate about their subject, titles in the series bring together science, art, literature, mythology, religion and popular culture, exploring and explaining the planet we inhabit in new and exciting ways.

Series editor: Daniel Allen

In the same series
Air Peter Adey
Desert Roslynn D. Haynes
Earthquake Andrew Robinson
Fire Stephen J. Pyne
Flood John Withington
Islands Stephen A. Royle
Moon Edgar Williams
Tsunami Richard Hamblyn
Volcano James Hamilton
Waterfall Brian J. Hudson

Moon

Edgar Williams

REAKTION BOOKS

To Royden

Published by
Reaktion Books Ltd
33 Great Sutton Street
London EC1V ODX, UK
www.reaktionbooks.co.uk

First published 2014

Printed and bound in China

A catalogue record for this book is available from the British Library

ISBN 978 1 78023 281 2

CONTENTS

Introduction 7

1 Lunar Shadows 11

2 Time and Motion 38

3 Stories and Legends 68

4 Man and the Moon 89

5 Lunar Art and Literature 107

6 One Small Step for a Man,
  One Giant Leap for Mankind 147

7 Legacy 172

  Timeline for Moon Landings 178

REFERENCES 180
SELECT BIBLIOGRAPHY 188
ASSOCIATIONS AND WEBSITES 189
ACKNOWLEDGEMENTS 191
PHOTO ACKNOWLEDGEMENTS 192
INDEX 194

Introduction

Ever since humans evolved from their earliest origins on the African plains several million years ago, the Moon has hung in the sky unchanged, passing monthly and unerringly through a cycle – at first a totally full, silvery-white orb lighting up the night sky, then briefly disappearing altogether before returning to its original glory.

The story of the Moon starts before life appeared on Earth. The Moon and Earth share a cataclysmic beginning in the early solar system around four billion years ago. Before the Moon was born, the newly formed Earth was still in its early throes of cooling down and solidifying when it was hit hard by another proto-planet of a similar size and was smashed apart by the huge impact. When the material reformed, the Earth and Moon were born largely as we know them today. For another two billion years the Earth and its offspring continued to be shaped by internal and external forces as debris rained down from space and volcanic action sculpted the surface through the ejection of copious quantities of lava and ash. With time, as the solar system slowly matured, the Moon settled into orbiting the Earth, seemingly content with its lot. The Earth, however, continued to develop. Because of its larger mass it held its atmosphere, and after gaining the lion's share of material during the collision was able to retain water that had rained down on its surface from stray comets, and through volcanism and plate tectonics constantly re-sculpt its surface, eventually allowing life to evolve. This life in turn altered the atmosphere and slowly human beings began their

rise to consciousness and ultimately their ambition to conquer the globe, which through the millennia had acted like clockwork, giving humankind the ability to organize time, allowing nations to rise and civilizations to flourish.

Before the Moon landings, the astronomer Patrick Moore (1923–2012) said, 'As the years passed in their millions, the fury died down, until at last the roar of the volcano was only occasional; and about the time the first sea-creatures appeared in the warm oceans of our own world, the Moon subsided into its long sleep – a sleep from which it will only be awakened by the coming of man.'[1]

In early antiquity, from the seventh to the eighth century BC, around the world and particularly in Asia and Europe, the Moon was worshipped as a deity along with many other gods and often paired with the Sun, sharing the sky as a sibling or loved one. The coming and going of the Moon as it arced across the sky traced a regular monthly journey, which to the ancients represented the cycle of life, death and resurrection. To these early observers, the Moon was mysterious and seemed to influence the weather and it soon became associated with water and dampness.

By late antiquity, as the Roman Empire established itself in Constantinople after the sixth century AD, and with the adoption of monotheism across Europe and the Middle East, the Moon, although still important to religious practice, had been demoted to a mere celestial body. Persian, Arab, Greek and Roman astronomers, having observed the Moon for centuries, now tracked and predicted its every move, phase and eclipse, all with quite amazing accuracy. This led to using each lunation (the time between each new Moon) as a unit of time – a month – allowing the dates of religious festivals such as Easter and Ramadan to be celebrated together across the vast Persian, Roman and Arab empires.

Even though the Moon lost its status as a deity, the idea that it influenced our psyche did not wane and many believed that madness was associated with the Moon. Excessive exposure to moonlight in the womb, at birth or while asleep was considered sinister and those under such lunar influence were considered to be suffering from lunacy, a particular affliction that was episodic

The Man in the Moon, a benign modern symbol of man's link with nature and the sky above.

in its nature and likely to occur at night (we now know that this condition was epilepsy). The feeling that there is a subtle lunar influence on our everyday living still abides today, and it continues to influence healthcare practice the world over.

The Moon has played a huge role in influencing our cultural history and predominates in all art forms, visual and written. To poets and playwrights the Moon has served as both an object of love and an object of sinister intent. It provides a mirror to reflect the poetic imagination. The twelfth-century Japanese poet, Kojiju wrote, 'Merely to know / The Flawless Moon dwells pure / In the human heart.'[2]

After the 1969 Moon landings, Selene, the moon goddess of the ancient Greeks, was thought dead and lifeless, a mere geological museum. Today that view has changed and we now know how vital the Moon is to the Earth's well-being, for without the lunar influence on the tides the Earth's spin would become ever more chaotic. Recent robotic probing of the Moon has shown that it contains water and minerals, giving nations across the world an urge to find out more. We are on the verge of witnessing a new era of lunar exploration, where the Moon is no longer Earth's poor offspring but a valued and respected neighbour.

This journey of discovery is described in the chapters that follow, first describing the physical Moon and then the metaphorical and metaphysical Moon, and its influence on our beliefs and cultures. The ever-present Moon has fascinated humankind throughout time and its appearance in world art and literature is described. We end our lunar journey with the Apollo Moon landings and their legacy.

1 Lunar Shadows

The Moon and Earth share a common origin in the solar system, which explains many of the similarities between lunar and terrestrial geology. Despite this the Moon has its own separate geological history, which includes the formation of mountain ranges, flat plains and basins. Added to this mix is the accumulated detritus from aeons of meteor strikes and subsequent crater formation, which means the mineral content of the Moon is complex. While its gravity and lack of atmosphere have meant reduced content of the lighter and more volatile compounds, the recent discovery of methane, helium 3 and frozen water in the dark fringes of craters in the polar regions has provided an exciting stimulus for a return to the Moon.

Our solar system formed 4.6 billion years ago, although no one is exactly sure how. While it was once thought to be unique, we now know the galaxy and universe abounds with stars and their planetary systems. The consensus of opinion is that the origins of our Sun, Earth, Moon and planets go back to a primordial cloud of gas and dust. At first the cloud was just a simple uniform entity, but over time, and through the fundamental interaction between mass and the force of gravity, inhomogeneous regions formed. The 'lumpy' cloud gradually formed into a spinning disc from which the Sun and planets eventually coalesced. In the disc, hydrogen, the lighter and commonest element, gravitated towards the centre, eventually becoming sufficiently compressed and hot to initiate nuclear fusion, creating a star, our Sun. The dust and lighter elements formed the inner rocky planets while the remaining gas and water ice formed the outer gaseous planets and comets. This evolution proceeded through the collisions within the clouds of trillions upon trillions of small particles, which gradually accreted into bigger and bigger bodies as collision followed collision. Then as the temperature around the primordial Sun cooled, the rocky bodies began to solidify, ultimately leaving us with the solar system as we know it today.

Our solar system has eight planets. The innermost and closest to the Sun, Mercury (diameter 4,880 km), circulates in an eccentric orbit ranging from 46 to 70 million km from the Sun. Mercury is a small, rocky planet, with a hot, rough surface, no atmosphere and an appearance similar to that of the Moon. The next, Venus, is a large rocky planet (diameter 12,000 km), circulating at 108 million km from the Sun. Venus has an atmosphere but not one that is favourable to life. Consisting of 97 per cent carbon dioxide and coupled with Venusian daytime temperatures of 462°c, it has one of the fiercest and most hostile environments in our solar system. While the two innermost planets are moonless, Earth (diameter 12,756 km), the third planet from the Sun and orbiting at approximately 150 million km, has an orbiting partner – the Moon. The Earth is at a sufficient distance from the Sun to achieve a surface temperature that is neither too cold nor too hot (around 14°c), to ensure that most of its water remains as liquid – essential for sustaining organic life. This is known as the 'Goldilocks' zone. Next is the red planet, Mars. At 228 million km from the Sun, it is rocky like Earth but smaller and devoid of any complex life. Mars has two odd-shaped moons, Phobos and Deimos.

The solar system changes beyond Mars and after the asteroid belt, a region of space filled with thousands of rocky bodies, ice and gas predominate over dust. Jupiter, the first gas giant, is of immense size: 1,321 times bigger than Earth and a massive 778 million km from the Sun. Saturn, Uranus and Neptune follow, before we reach the Kuiper belt and finally the Oort cloud. This is a belt like the asteroid belt of small, randomly distributed frozen bodies, which includes Pluto. No longer the ninth planet, Pluto has joined a family of dwarf planets.

These outer planets have many natural satellites or moons. While the planets and dwarf planets number fewer than twenty, there are more than 100 moons in the solar system. Jupiter and Saturn have the most, around 84 between them, and even diminutive Pluto has five. Earth's one large natural satellite, the Moon, is not the largest in the solar system. This title goes to the Jovian moon, Ganymede, which with a diameter of 5,268 km is bigger than Mercury. Two other Jovian moons are worth a mention: Europa and Io. Europa is slightly smaller than the Moon and is encased in water ice around 100 km thick, which covers a vast ocean of liquid water beneath; it is a moon still waiting to give up its many secrets. Io is heated by its intimate orbit around Jupiter, resulting in one of the most volcanically active bodies in the solar system. Titan, a moon of Saturn, the second largest moon in the solar system and larger than Mercury, has an atmosphere and seasonal weather patterns. Being cold, its clouds are not made of water but the hydrocarbons, methane and ethane. Thus the Titan landscape is shaped and weathered by a hydrocarbon rainfall, and covered in lifeless lakes and rivers.

Although our own Moon seems rather dull in comparison, we do know a lot more about our own neighbour. We know that it has remained virtually unchanged since its formation 4.5 billion years ago, and that it is largely made up of granite-like rock. The Moon is only a quarter the size of the Earth, with a diameter of 3,476 km, giving it a surface area about the size of Africa. In terms of mass and volume the Moon is even less substantial, with a density of 3.34 g per ml, its volume 2 per cent and mass 1 per cent that of the Earth. This gives the Moon a surface gravity one-sixth the strength of the Earth's gravity.

The origin of our Moon has occupied the thoughts of people since the beginning of time, and is reflected in the folklore of many of the world's religious texts, legends and myths. Most of these early views give the Moon an anthropomorphic or spiritual origin, often ascribing it the status of a deity. In the book of Genesis, the Moon is created, along with the Sun and the stars, on the fourth day: 'And God made two great lights, the greater light to rule the day, and the lesser light to rule the night: He made the

A tin-glaze plate of 1733 depicting God creating the Earth and Moon, a design taken from 'Raphael's Bible', a series of early 16th-century frescoes by Raphael decorating the Vatican's Loggia.

stars also' (Genesis 1:16). This view was largely unchallenged until the eighteenth century, when the science of terrestrial geology was born. Theories concerning the Moon's origins were soon developed by astronomers, naturalists and scientists.

One of the earliest scientific theories about the Moon's origins, and one of the most benign, was suggested by the work of William Thomson, first Baron Kelvin (1824–1907), who postulated that in the far and distant past the Earth and Moon had formed out of a dust cloud alongside one another and had since existed as we largely see them today. This simple status quo theory soon began to lose its appeal when careful measurements of the lunar orbit showed that it was not steady but was changing with time. To account for an orbit that was slowly spiralling away from the Earth, another possible genesis was suggested in which the Moon formed independently elsewhere in the solar system and, after drifting in an eccentric orbit, was captured by the Earth's gravity as it crossed the Earth's orbit. This

Ganymede, the largest moon in the solar system, first observed by Galileo Galilei in 1610.

would explain why, with a diameter one-quarter of that of the Earth, the Moon is unusually large in relation to its companion (the majority of moons in the solar system are many times smaller than their associated planets). If this were the case such an alien Moon would have a different geological history and structure to Earth, reflecting the region of the solar system in which it was formed. This idea was refuted when it was found that the Moon's geology and planetary characteristics closely represented the Earth (facts amazingly enough gleaned solely by telescopic observation of its surface and motion). Another theory, the fission theory of 1878, was championed by the astronomer and mathematician George Darwin (1845–1912), the fifth son of Charles Darwin, who postulated that the Moon was created when it was torn away from the early Earth by centrifugal force. He theorized that as the molten Earth span rapidly on its axis it formed a dumb-bell shape. Eventually the neck of the dumb-bell

Francis Danby, *Mt Etna at Sunset with a Crescent Moon, c.* 1829.

broke and the two bodies – Earth and Moon – separated. The effect of this traumatic separation created the Pacific basin, a huge circular depression in the globe's surface, while at the same time tearing apart the large land mass on the other side of the world (Africa and South America). This accounted for the Moon's similar structure to the Earth and also conveniently helped explain the interlocking shape of South America and Africa. This theory was particularly popular as it explained the formation of the Pacific and Atlantic oceans and the distribution of the present terrestrial geography. It finally fell out of favour as it was realized that the Earth could never have spun fast enough to generate the centrifugal forces needed to fling the mass from the surface into orbit. With the realization in the 1980s that some meteorites were of lunar origin and their composition slightly different from similar rocks on Earth, the Moon capture theory championed particularly by the Nobel laureate and chemist Harold Urey (1893–1981) in the 1950s returned to favour.

Towards the end of the twentieth century a more advanced and informed theory developed, postulating that the Moon was born following a cataclysmic collision between the early Earth and another planetary object with an irregular orbit that crossed Earth's orbit. This collision would have occurred around 4.5 billion years ago, when the smaller proto-earth was still largely molten and not quite yet solidified. The colliding object would have been roughly the size of Mars, and has been given the name Theia, after the mother of Selene, the Greek moon goddess. The collision was so cataclysmic that it knocked the Earth off its axis, resulting in the current 23.5° tilt we see today. The tilt, now stabilized by the Moon, gives us our seasons during each orbit of the Sun. The impact was so catastrophic that the Earth fractured and any early atmosphere lost, with Theia being largely destroyed. The majority of the pulverized Theia mixed with the Earth and the remaining ejected debris formed a hot disc of vapour and material around the planet, much like the rings of Saturn. Then over many centuries some of the orbiting debris fell to Earth and mixed with the cooling surface while the remainder coalesced to form the orbiting Moon. This created a Moon with a similar geological composition to that of the Earth, with Theia's fingerprint being completely obliterated in the process.[1]

This theory has some limitations. While it requires the total destruction of Theia (as the mythological Theia sacrificed herself for her daughter, Selene), it also requires that the orbiting ejecta would have to remain very hot for a long time to allow sufficient mixing to occur. It is very difficult to achieve this level of heating from a head-on collision. Thus a more recent theory suggests that the collision was not head-on, but was more like a high-speed glancing blow. This sort of 'hit and run' collision creates the heat required, and if this theory is correct then somewhere in the solar system a remnant of Theia may still survive.[2]

The new Moon would have formed close to the Earth with both spinning faster than they do now, resulting in the terrestrial day being only a few hours long. The tidal effect of gravity between the two bodies eventually slowed the rotation of both Earth and Moon, and pushed the Moon into an orbit of ever increasing diameter, eventually locking its rotation to ours. Viewed from the Earth, the Moon appears to have lost its rotation altogether, always presenting the same face.

The collision with Theia left the Earth with a heavy molten-iron, radioactive core, giving it the ability to generate heat and a protective magnetic field. The Moon was left with only a small molten-iron core and a very weak magnetic field. Since then the dynamic interaction between the Earth's liquid magma core and its hard outer layer, the mantle, ensures that the Earth's surface has been in continuous flux and the surface is always being subsumed or reformed through volcanic eruptions somewhere in the world. While orbiting debris would have rained down onto the surface of the cooling Earth and Moon for millennia, there is scant evidence of this debris on the Earth, as it has been lost in the continual geological reshaping and later weathering of the surface. Remarkably on the Moon, a large proportion of the original surface remains just as it formed those many billions of years ago. While the surface is not flat and has undergone some volcanism of its own, there are plenty of valleys and mountains left over from this era. While moonquakes do occur they do not alter the surface and are small – more like tremors. The Moon's surface therefore provides a record not only of the Moon of billions of years ago, but of all the impacts that have rained down on its surface throughout its lifetime.

The surface of the early Moon would have been molten as it cooled, producing smooth basaltic lava plains and basins. Similar features, but formed later (up to about one billion years ago), can be seen from the Earth as dark patches, which the early observers thought were large tracts of water, thus naming them oceans or seas (or Maria). On the near side they cover about 30 per cent of the surface while on the far side they are rare and the terrain here is more mountainous.[3] It is thought that while the surface was still molten a thin crust formed, and volcanoes continued to spew out lava as breaks and cracks in the surface allowed the more molten magma just below to vent and escape. Once the Moon cooled, all these geological processes ceased, creating a non-spherical Moon. To complicate matters, during this time the Moon was constantly bombarded by asteroids and comets from both the inner and outer solar system. These formed impact craters, of which there are an estimated 300,000. Along with craters the impacts also formed hills and mountains at their centres. These features formed in a few momentous minutes or hours rather than the many millennia on Earth through the process of plate tectonics. The size and rate of the impacts varied over time. One of the earliest records of an impact on the Moon is from 4.3 billion years ago, near the south pole. The object, which was a few hundred metres across and struck at a low angle, created the South Pole-Aitken basin. Being principally on the far side of the Moon, it was not known until the 1960s when probes were first sent around the Moon. At 2,500 km across and in places 6 km deep, it is not only the oldest but the largest lunar basin (also the largest in the solar system).

Then around 3.9 billion years ago there was an intense bombardment, which peppered the surface with thousands of craters. Even now there are around 1,700 large craters (over 20 km in diameter) on the Moon dating from this time. The Late Heavy Bombardment, as this period is known, occurred throughout the

Artist's impression of the Giant Impact hypothesis, a collision between the Earth and a massive object, debris from which formed the Moon.

inner solar system, and the Earth, with a diameter four times the size of the Moon's, would have received a greater bombardment and be pockmarked with ten times as many craters. Again weathering and tectonic activity have obliterated all evidence of this bombardment on Earth. Our own origins date back to the end of this bombardment, and the earliest fossils date to then. It seems that the water released from the Earth's crust during these impacts and the global conditions engendered by the large influx of material and energy from space, including water and organic material, created the conditions needed to accelerate the evolution of life.[4] Thus the Moon and Earth shared their infancy in a solar system full of hazards and uncertainty.

Between 3.8 to one billion years ago there was intermediate cratering resulting from thousands of small impacts. After this the numbers of large impacts were much fewer. Notable impacts formed the Copernicus crater 800 million years ago and the Tycho crater 108 million years ago. Tycho is 82 km in diameter and located near the South Pole. It was first scientifically described in 1645 and later named after the Danish astronomer Tycho Brahe (1546–1601). Being so young it has not been changed by lava flows or degraded by further large impacts and thus has a smooth interior, making it reflective (it has a high albedo) and easily visible from Earth (it can be seen with the naked eye). Being a relative youngster, the 'splash' of the molten rock created during the impact can still be seen as a series of rays radiating away from Tycho's centre. These rays are hundreds of kilometres long and are visible with binoculars. More recent and much closer observations made by orbiting craft show that within the 4.7-km-deep crater there is a central peak that is 2,200 m high. This huge mountain was formed in an instant by the impact and remained frozen in time, un-weathered for millennia. Unusually for the Moon it has steep sides. Recently, an even more extraordinary discovery was made: a more detailed close-up of the peak revealed that sitting atop this

The Aitken Crater
on the northern edge
of the South Pole
Aitken Basin.

huge mountain is a single boulder
120 m across. It is not known how
it got there. Notably this location is
close to the site of a fictional alien
monolith discovered in *2001: A Space
Odyssey* (1968), a science fiction novel
by Arthur C. Clarke (1917–2008),
who would have known nothing
about this extraordinary rock.

There are about eighteen mountain ranges on the Moon, with familiar names such as Montes Alpes and
Montes Pyrenaeus (named after the
European Alps and Pyrenees mountain ranges respectively). Several of
their peaks reach over 3,000 m in
height, the highest being Mons
Huygens at 4,700 m. The Moon's
surface is cut by valleys, the largest
being Vallis Snellius at around 600
km in length.

The lunar surface has other
notable features such as rilles, a
word derived from the German for
'groove'. Three types of rille have
been identified. There are sinuous
ones, which are ribbon-like and look like dried-up river valleys, but were formed by flowing molten lava and old collapsed lava tubes. Vallis Schröteri in the Oceanus Procellarum is the largest sinuous rille. The second type, the arcuate rille, is curved and circular in shape. These form around the edge of an impact crater and result from slumping crater rims. The third type are long linear grooves of sunken crust that run between adjacent factures on the surface, equivalent on Earth to grabens. Rimae Pettit at 450 km is the largest such rille on the lunar surface. Other strange features are the so-called lunar swirls, an example being the Reiner Gamma swirl named after the nearby Reiner crater.[5]

The Copernicus crater.

An oblique view
of Rima Ariadaeus,
photographed by
Apollo 10 astronauts
in May 1969.

These swirls are light patches and are not associated with any other geological features. Swirls, which have always interested the earliest explorers, are related to magnetic anomalies in the Moon's structure.

The Moon lacks an atmosphere but is surrounded by a diffuse shell of gas, or an exosphere, which, despite being very thin, contains a surprisingly diverse range of gases including helium, nitrogen, argon, oxygen and radioactive radon and polonium. The exosphere is so diffuse because the gases constantly move out into space. The exosphere is not lost, however, because there is a constant efflux of gas seeping out from the Moon's interior, created by radioactive decay. The strong ultraviolet light from the Sun causes the gases to become electrically charged, making the sky glow at night.

Because there is so little atmosphere the lunar sky appears black, in contrast to Earth where the sky is given its blue colour by water vapour via reflected sunlight. This unfiltered lunar light is brighter than our sunlight and outshines most celestial objects. When you look at the Apollo images the sky is completely black. The astronauts only went walkabout during the lunar daytime, in the early morning when it was not too hot. If they had visited during the night the blackness would have been absolute, as Earthshine is not that bright (especially if there was a 'new Earth' phase). This lack of celestial objects in the background was quickly ceased upon by the conspiracy theorists who believed that the Moon landings were faked and took place in a film studio somewhere in the u.s. The loose nature of the dusty surface, the regolith, causes a diffuse reflection, so the ambient shadows and light are unlike reflected sunlight on Earth.

Until recently, the Moon was considered to be completely devoid of water because of its cataclysmic birth and lack of atmosphere. Geologists were so sure of this that when the Soviets examined the first ever lunar samples and found that they contained a minute trace of water, it was incorrectly dismissed as Earthly contamination. Most of the recent remote sensing satellites sent into lunar orbit have now confirmed the ubiquitous presence of water, albeit in trace quantities. Some of this water has come from the Sun, formed from hydrogen carried by the solar wind reacting with the oxygen in the lunar soil to create water-based compounds. This process might not be confined just to the Moon – all the inner bodies of our solar system, such as Mercury, could accumulate water in this way. More exciting, though, is the fact that water ice has been detected deep inside the craters in the Moon's polar regions, which are never exposed to sunlight. The Hermite crater at the South Pole has the record of being the coldest place in the solar system, colder even than Pluto. It is permanently in shadow and its lowest recorded temperature is -248°c. A potnetially abundant water deposit is thought to lie within the three-billion-year-old Shackleton crater near the South Pole. The crater, named after the Antarctic explorer Ernest Shackleton, is huge, being 4 km deep and 21 km across. Due to its near vertical orientation, the rim is in permanent sunlight but its steep sides and bottom never see the Sun, and have remained deeply frozen for millennia, a sunless abyss. The soil may contain around 5 per cent water, which in terrestrial terms is dry but in lunar terms is a great deal, and given the size of the crater would, if extracted, provide a fair cornucopia of water for any lunar visitors.[6] Recent research has shown that there may be around 100 craters that are permanently in the shade and therefore might contain water ice.[7] In contrast to these permanently shaded craters there is a small area in the North Pole where the craters are in near permanent sunlight (96 per cent), making it hotter than most other polar regions.

The surface temperature of the Moon was first measured by Lawrence Parsons, 4th Earl of Rosse (1840–1908), in the 1880s, using a thermocouple (a thermometer-like device) attached to his telescope.[8] After the Second World War, with the invention of radar and radio astronomy, the Moon was observed to emit a weak radio signal as the surface was heated by the Sun. Measuring this signal allowed scientists to work out the lunar surface temperature. Next, by bouncing radar off the lunar surface, scientists were able to calculate the change in surface temperature over the lunar day and, more importantly, the composition of the surface. During the day the surface temperature rises to a hot 127°c while

This depiction of the bulk density of the lunar highlands was made by means of gravity data from NASA's GRAIL mission in 2012 and topography data from NASA's Lunar Reconnaissance Orbiter. Red indicates higher than average densities and blue lower ones.

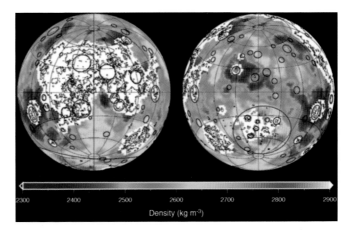

Density (kg m⁻³)

at night it plummets to -173°c, cold enough to freeze most gasses into solid ice.

The Victorians argued that without any wind or rain to weather the surface, over aeons the surface would be covered in a light dust created by mechanical wear and tear of the rocks, possibly many metres thick. This would have made lunar landings more difficult. However, the radar results showed that the surface was covered in fine dust but this was only a few centimetres deep. Below this was a layer of gravel and porous rock.[9] Recently, as part of the GRAIL mission, two probes mapping the Moon's gravitational field have found that during its first billion years the plethora of impacts pulverized the lunar surface and crust beneath it, and it is quite thin at only 34–43 km thick.[10] The 'lunar soil' is composed mainly of oxygen silicon, iron, calcium, aluminium and magnesium, although rarer elements like titanium and helium3 are in greater abundance than on Earth. On Earth, helium3 is measured in grams, while its reserves on the Moon are estimated at half a million tons. One incentive for our return to the Moon may be to mine its great mineral wealth.

Moon dust is toxic to man, being very fine and, because it has not undergone aqueous weathering, very sharp-edged. The dust formed after meteorites vaporized on impact and the resulting high temperatures produced a glassy and gritty ash. This will damage the lungs if inhaled, causing silicosis, just as volcanic ash

does on Earth. The dust would irritate the skin and be particularly damaging to the cornea. The amount of radioactivity associated with the dust may cause cancer if it is ingested or inhaled. Thus when humans travel to the Moon they need to keep dust at bay. In addition, its effect on instruments with moving parts can also be disastrous.[11]

A bootprint left on the Moon on 20 July 1969 by one of the Apollo 11 astronauts.

Mapping the Moon

Although the astronomer and mathematician Thomas Harriot
(1560–1621) drew the crescent Moon in 1605 using his primitive
telescope, it was not until Galileo Galilei (1564–1642) turned
his telescope on the Moon in May 1609 that the mapping of
the lunar surface began. Galileo's observation that the Moon was
a three-dimensional object, a sphere rather than a disc, created
as much of an impact as the discovery of America had a few
centuries earlier. His observations revealed that the Moon was
a new world with mountains, valleys and plains hundreds of
kilometres across. He even calculated the height of the mountain by measuring the lengths of their shadows when crossed
by the terminator (the moving circular boundary line demark-
ing the lit and unlit areas of the lunar surface of the waxing
or waning Moon when observed from Earth). It was easy to
speculate that the Moon contained life, but difficult to prove.[12]
This upset many of his contemporaries who thought the Moon
was perfectly smooth and the spots on its surface were reflections
of our own mountains, or shadows of opaque bodies floating
between the Moon and Sun.

The first realistic lunar map with names was made and
published by Michel Florent van Langren (1598–1675) in 1645,
and this was followed in 1647 by *Selenographia* by the amateur
Polish astronomer Johannes Hevelius (1611–1687), who had
built his own observatory on the roof of his house in Gdansk.
Although the map was small at about 30 cm in diameter, he
labelled the features using names from classical mythology and
from geographical features on Earth. His now Latinized nomenclature is still evident, for example in the mountain range the
Alps or Montes Alpes. Upon his death, the copper printing plate
of the map was destroyed and ended up as a teapot and so his
map was lost to posterity.[13] The next significant map was published in the *Almagestum novum* in 1651 by the Italian astronomer
and Catholic priest Giovanni Riccioli (1598–1671) and drawn
by Francesco Grimaldi (1618–1663). This map was less accurate
than that of Hevelius, but is remembered because of Riccioli's

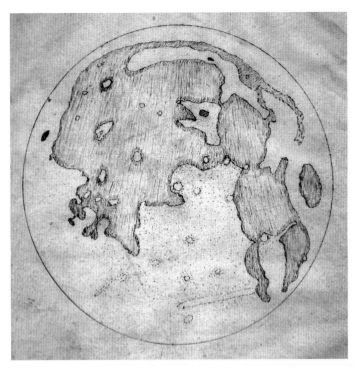

An early copy of one of Thomas Harriot's Moon maps of 1609–13, with the Copernicus Crater shown at upper left.

Galileo's first sketches of the Moon's surface based on his early telescopic observations.

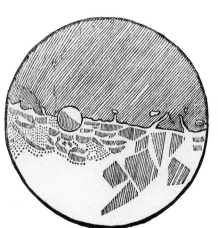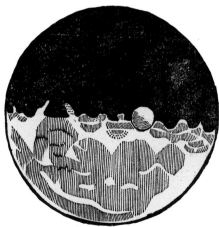

Giovanni Riccioli's map of 1651 named the dark patches on the Moon surface as seas, a system still in use today.

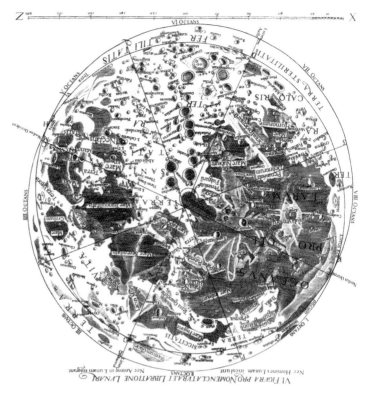

innovations of Latinized nomenclature for large and small features such as craters, and naming of them after contemporary and past eminent scientists and philosophers, ensuring along the way that a crater was named after himself and his colleague Grimaldi. His naming system is still in use today.

The early observers thought the large magma plains were huge oceans, just like those covering the Earth. On the surface there are around 22 named seas and one ocean, Oceanus Procellarum (the Ocean of Storms). The seas have a range of names from the mundane Mare Australe (Southern Sea) to the bizarre Mare Fecunditatis (Sea of Fertility). The best known is the Mare Tranquillitatis, or Sea of Tranquillity, made famous by the first manned landing on the Moon by the Apollo 11 mission in 1969. The two eyes of the 'Man in the Moon's' face are seas, the left eye the Mare Serenitatis (Sea of Serenity) and the right the Mare

Imbrium (the Sea of Showers or Rains). The mountain range
Montes Appeninus forms the bridge of his nose while his smiling
mouth is made up of a number of crater-pocked plains.

It was over a century before Johann Hieronymus Schröter
(1745–1816) produced a more detailed lunar topography, using
a 16.5 cm aperture reflector telescope. His work, strangely entitled *Selenotopographische Fragmente zur genauern Kenntniss
der Mondfläche* (Selenotopographic Fragments for More Exact
Knowledge of the Lunar Surface), was published in 1791. This
was followed in 1836 by *Mappa Selenographica*, which contained
the most detailed maps yet, produced by the German banker
and amateur astronomer Wilhelm Beer (1797–1850) and the
astronomer Johann Heinrich von Mädler (1794–1874). This
innovative work divided the Moon into quadrants and took
several years of close telescopic scrutiny to produce. It was followed a year later by a further publication which provided in
great detail the diameters of around 150 craters and the heights
of around 830 lunar mountains. By the end of the nineteenth
century a revolution in photography had occurred and photographic glass plates had become sufficiently sensitive to low light
to allow detailed lunar images to be made without blurring. The
revolution in hi-fidelity lunar photography began Dr John
William Draper (1811–1882) from New York in 1840 and was
perfected by 1895 by the astronomers Maurice Loewy (1833–
1907) and Pierre Puiseux (1855–1928) at the Paris Observatory,
who produced a four-volume atlas (*Photographique de la lune*,
1896–1910) containing over 6,000 photographs of the lunar
surface. For this contribution a crater in the Mare Humorum
was named in Puiseux's honour. After this, professional interest in Moon topography began to wane and in the twentieth
century lunar mapping became the pursuit of the amateur
astronomer. However, there was still a lot of detail to be seen
using the small telescope as Patrick Moore, the great amateur
advocate of lunar observation, found. Using a telescope and
observing the passage of the terminator crossing the surface,
especially when the Sun was low on the lunar horizon, provided
ample opportunity to observe new and sometimes transient

features. Indeed Patrick Moore's maps were used by the USA's National Aeronautics and Space Agency (NASA) to help plan the manned Apollo lunar landings.

Show someone an image or map of the Moon and they seldom know which way is north. To make matters more opaque, the Moon viewed through a telescope appears inverted, so astronomical images are often shown with the South Pole at the top. For the uninitiated, a cursory look at the Moon shows nothing more than a random splattering of white and grey blobs. The sharp-eyed will notice some features like the crater Tycho. Others will squint and try to orient the map so that they can see the man in the Moon's face. Conversely, almost anyone can recognize the shape of the continents and oceans on a cloudless image of the Earth. To complicate matters further, someone observing the Moon from the northern hemisphere sees its North Pole at the top whereas an observer in the southern hemisphere will see the Moon inverted with the North Pole at the bottom. In the

A late 18th-century engraving detailing seas and craters that gives a more modern impression of the Moon's surface.

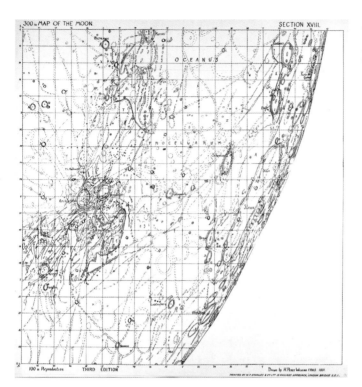

A section of a highly detailed Moon map drawn by H. Percy Wilkins in 1951.

Tropics the Moon is sometimes south and sometimes north of the equator so its face switches between the two.[14]

On Earth, craters (from the Greek term for 'cup') were closely associated with volcanoes, a term coined by Aristotle and referring to the cup shape of the craters found inside the summits of a typical conical volcano. Good examples are Vesuvius in Italy and Fuji in Japan.[15] Thus lunar craters were assumed to be the mouths of great volcanoes seen from above. While some scientists like Robert Hooke (1635–1703) theorized that craters were caused by impact, this was not widely accepted, as space was considered to be empty (apart from stars, planets and comets) and so devoid of any material to impact with the Moon. At that time there were no known impact craters on the Earth. It was assumed that the Moon was protected by its atmosphere, much as the Earth is today.[16] However, as telescopes grew bigger and more powerful, so did the certainty that craters were volcanoes. In 1787 the

A full Moon over Namibia, depicting a lopsided lunar landscape.

astronomer William Herschel (1738–1822) even reported seeing active volcanoes – not observing actual flames but interpreting the brighter crater floor as volcanic activity, describing the glow as resembling 'a small piece of burning charcoal, when it is covered by white ash'.[17] Herschel and many of his contemporaries were also convinced that the Moon was inhabited. The astronomer Johann Hieronymus Schröter claimed to have observed green fields on the Moon.[18] The belief that the crater rims were mountainous like Vesuvius was further championed by two British amateur astronomers, James Nasmyth (1808–1890) and James Carpenter (1840–1899), whose book, *The Moon: Considered as a Planet, a World and a Satellite* (1874), showed detailed plaster models of the lunar surface photographed with low-angle lighting. These techniques produced details that looked similar to those observed on the lunar surface. The authors have craters named after them, Nasmyth in the lunar south and Carpenter in the north. Ironically the Nasmyth crater is flat and is clearly an impact crater with no 'volcanic' peak.

A year earlier another English astronomer, Richard Anthony Proctor (1837–1888), who wrote *The Moon, Her Motions, Aspect, Scenery and Physical Condition*, was one of the first to advocate the collision hypothesis. The idea was largely forgotten on the publication of the Nasmyth and Carpenter book. The opponents of the impact theory wondered how this method of formation could create such perfectly circular craters, being of the opinion that impacting bodies would create smeared craters, especially as they would have hit the surface at an oblique angle.

Another astronomer, William Henry Pickering (1858–1938), in his book *The Moon: A Summary of the Recent Advances in Knowledge of our Satellite with a Complete Photographic Atlas* (1904), thought the shape of the Moon's surface could be explained by snow and ice cover. He believed craters were not volcanic but were caused by steam explosions erupting from beneath the ice. Up to now it had been only professional astronomers who seemed able to pronounce on Moon formations, but once geologists' views were listened to the views began to change. In 1921 Alfred Wegener (1880–1930), the geologist famous for his work on

One of James Nasmyth's plaster models of the lunar surface, 1863.

35

continental drift, argued that craters were caused by impact. The circular bomb and shell craters created across northern France during the First World War indicated that high-impact craters were round, whatever the angle of impact. It also showed how a central peak could form following impact. Many of the Moon's craters, such as Tycho, have these peaks. As this point two camps of scientists developed – the traditionalists, who supported the theory of volcanism, and the physicists and geologists, who believed in the impact theory. The dominance of meteoric impact as an explanation of the Moon's features was secured in 1949, when Ralph Belknap Baldwin (1912–2010) published *The Face of the Moon*. By now evidence for terrestrial craters and impacts was mounting, including the massive Chicxulub impact structure in the Yucatán Peninsula in Mexico, which is 180 km across. Despite this a few astronomers held on to the volcanic hypothesis. Patrick Moore in his *Guide to the Moon* still championed the volcano theory, and even in 1965 the British astronomer Gilbert Fielder was a strong advocate. It was not until after 1969, with the return of lunar samples directly from the Moon, that the volcanic theory was finally buried. Fortunately we are living in a time when meteoric impacts are rare and it is easy to see why it could seem

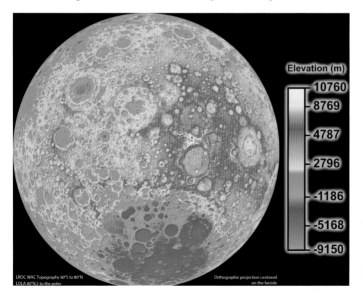

Elevation (m)

10760

8769

4787

2796

-1186

-5168

-9150

LROC WAC Topography 80°S to 80°N
LOLA 80°N,S to the poles

Orthographic projection centered
on the farside

This map created by NASA in 2011 is colour coded to show the topography of the lunar surface.

unlikely that meteors could influence the development of the Moon's or the Earth's surface or even alter our biological environment. Furthermore, on Earth we are surrounded by active volcanism all the time, so it is understandable that this theory dominated for so long.[19]

Strangely enough, the lunar far side is different to its near side: it is more mountainous and has suffered more hits from meteorites. Facing away from the Earth, it has absorbed many more interplanetary impacts, protecting both the Moon's near side and the Earth.[20]

Since we have visited the Moon and sent a plethora of spacecraft to measure and photograph it from every conceivable angle, we now know her features at the most intimate level. On Earth it is the movement of the Moon rather than its composition or topology that has most affected humankind and, as we shall see in the next chapter, the interaction between the Sun, Earth and Moon is not a simple one.

2 Time and Motion

From the earliest of times humankind has found the phases of the Moon easy to observe, and this has provided one of the earliest systems for measuring time. The lunar cycle, the time in which the Moon passes through all its phases (roughly a month), became important in the evolution of calendars in nearly all the ancient civilizations. The ancient Greeks used it to describe a dual four-year Olympiad consisting of two consecutive periods of 50 lunar cycles and 49 lunar cycles amounting in total to eight solar years. The lunar month also dictates the timing of religious festivals such as Easter, Ramadan and the Chinese New Year.

The Moon is the most prominent celestial object in the sky and, to the casual observer, appears at random times and positions, sometimes at dusk, sometimes late at night, and even during the day. This casual observer may even notice that on some nights the Moon never shows at all, and that it is usually not a complete disc, but only a part thereof, and sometimes a left-facing portion and sometimes right facing. The position of the Moon and its phase is dependent upon its position relative to the Earth and Sun and to some extent Venus. In fact, to account exactly for all the interactions between the Moon, Earth and Sun, and the different variations in its position, time and phase, requires a whole range of mathematical equations with up to 1,500 variables.

While the Moon orbits the Earth the duo orbit the Sun, with a complete orbit of the Earth around the Sun marking a year (known as a solar year), which takes 365 days or, to be precise, 365 days, five hours, 48 minutes and 48 seconds. Ignoring these

A full Moon on a clear night, shining in all its glory.

extra hours is not advisable, as this would soon make any calendar inaccurate. After a century it would be a day behind. During a solar year the Moon orbits the Earth around twelve times. These twelve lunations, marked as twelve months, are easy to observe and were used by the earliest civilizations to create the first primitive calendars. One of the earliest lunar calendars comes from the Aurignacian culture in Germany and is estimated to be 32,000 years old. These late Upper Palaeolithic people recorded the lunar cycle by creating a timeline of full and crescent Moon shapes on bone. Another example, painted 15,000 years ago on the Lascaux cave walls in France, shows a series of marks thought to represent the Moon going through its phases.[1] The Moon orbits the Earth in 29.53 days, which means that a simple lunar calendar of twelve months is also inaccurate and within a solar year would be out of sync by eleven days. A simple solution is to add twelve extra days to the calendar, something the Romans did by alternating between months of 30 and 31 days.[2]

By 3000 BC the Babylonians were able to calculate the movement of the Sun, Moon and planets, allowing them to produce accurate calendars with each new year, starting on the first new Moon after the spring equinox (very much like Easter). They did this by observing the quarters of the Moon. These became *shabbattu* and were later adopted into the Jewish calendar, becoming the Sabbath, a day occurring every seven days. In the Christian calendar this became a week, with Saturday and Sunday becoming the weekend.

Around the same period the Assyrians were also keen observers and noticed after years of careful record-keeping that

Bramantino's *The Crucifixion*, 1510–11, depicts Christ and the two thieves overlooked by the Sun and Moon.

there was a complex interaction between the orbits of the Moon and Earth which completed a cycle every 18.6 years. This cycle, now known as the metonic cycle, led to the early development of a twelve-month calendar that stayed synchronized with the solar year. Stonehenge, the Neolithic stone circle on Salisbury Plain in the UK, is thought to have two upright stones marking moonrise at the two extreme positions of this lunar cycle.

Even this lunisolar calendar was still not perfect and after a few centuries it needed correcting. It was the Romans and Julius Caesar who in 45 BC introduced a leap year every fourth year by adding an extra day. Even this introduction of the extra day (our 29 February), wasn't enough to keep the solar and lunar calendars aligned. Thus on 24 February 1582, Pope Gregory XIII initiated the Gregorian calendar. At this time the Julian calendar was ten days ahead. The Romans had assumed the year to be 365 days and six hours, which is around eleven minutes too long. Shortening the year by this amount has put the calendar on track and it is the one in use today. The reason the Catholic Church amended the calendar was because they relied on the spring equinox to define Easter. The new Gregorian calendar jumped these ten days to restart afresh and modified the quaternary rule, such that centenary years that were divisible by 100 were not leap years, except if they were also divisible by 400, which makes 2000 a leap year but not 2100. Since then the lunar and solar calendars have been synchronized, although the occasional leap second is still required as the Earth's rotation gradually slows.

The Church deemed Christ's Resurrection to have occurred during the Jewish Passover, which is fixed to the lunar calendar. Emperor Constantine (AD 325) wanted to find a unified date for its celebration, using a solar calendar, fixing Easter day to the vernal (spring) equinox on either 21 or 25 March. So he convened a Christian council in Nicaea, near Constantinople, which was attended by around 300 bishops (this was Christianity beginning to come together as a single religious universal – Catholic – church). At this point each bishop was free to follow his own views. The four Gospels are not clear on which Sunday after the

A 14th-century
enamelled copper
plaque showing
Christ crucified
among the Sun,
Moon and stars.

Passover feast the Resurrection occurred. Precise times and dates
were not recorded at the time, as it was a given that God was
timeless, but as the centuries passed, dates needed to be set in
order to celebrate saints' days and schedule Christian holidays.
The council agreed that 'by the unanimous judgement of all it
has been decided that the most holy festival of Easter should
be everywhere celebrated on one of the same day'.[3] This was to
be the first Sunday following the paschal full Moon, which is
the full Moon that falls on or after the vernal equinox. We still
follow this arrangement today.

　　As we have already seen, the pre-Julian calendar needed
additional days to remain synchronized, so months were paired,
with one of each pair giving a full Moon and the other a 'hollow'
Moon (hollow referring to the shape of the Moon during its

first quarter). Using this system extra days were inserted every three years. To decide on the time of appearance of the appropriate lunar phase the Romans used an official called a Pontifex to proclaim when the Moon became visible, thus signifying when each month began and whether the nones (the appearance of the Moon's first quarter as a crescent) of the month would fall on the fifth or seventh day. This system was soon dropped by the Romans, but it is still used in the Islamic calendar, and Ramadan, a month of fasting, officially begins when the first crescent of a new Moon is sighted and ends after the next new Moon. The beginning and end of Ramadan usually fall on single days, but the actual time varies across the world as lunar sighting depends on longitude or time of day and position in the sky. To help, a Saudi Arabian coordination committee of major mosques and Islamic centres officially announce the time of the crescent Moon's appearance. Being a lunar event, Ramadan begins about eleven days earlier each year, and thus occurs in every month of the year, taking 33 years to complete a cycle.

Christianity has driven the development of calendars away from the lunar calendar to the more accurate solar calendar, providing evidence of God's work and the taming of the universe. It

A lunar eclipse illustrated in Charles F. Blount's *The Beauty of the Heavens* (1845).

A digital montage showing
several stages of a lunar eclipse.

was therefore strange that later, when the heliocentric model of the universe was proposed, it was the Christian Church that opposed it most vehemently. Today, there are a few remnants of this link between the Moon and our calendar. For example, the French word for Monday, *Lundi*, derives from the goddess Luna. 'Monday' is derived from an old English word meaning Moon's day, naturally following on from Sun's day.

Lunar eclipses

In the seventeenth century the Moon played a central role in aiding Europe's influence around the globe. Navigation to far-off lands requires knowing where you are at any one time. At this time the latitude (your vertical position between the pole and equator) was measurable, but estimating longitude (your position west or east) was more difficult. Astronomers knew that the timing of predictable celestial events depended on your longitude. For example, observing the exact moment a lunar eclipse occurred or when a fixed star passed behind the Moon would allow longitude to be calculated. This was fine on land, where astronomers had access to accurate clocks, but at sea, and in unfamiliar regions of the world, accurate timekeeping was

A solar annular eclipse, where the Moon's passage between the Earth and Sun does not completely block out the Sun. From Blount's *The Beauty of the Heavens* (1845).

Swaine Monamy,
*Moonlight Scene:
Ships Saluting,*
c. 1800.

difficult. In Spain, Michel Florent van Langren (1598–1675), royal cosmographer and mathematician to Philip IV, was tasked with improving navigation methods to the Spanish New World. The king wanted an accurate map of the Moon so that it could be used to determine longitude at various locations in his empire. Longitude was to be estimated by comparing the difference in sunset and sunrise times of certain lunar features as they crossed the terminator. Unfortunately this proved inaccurate as a navigation method because of the lack of accurate portable timepieces. Van Langren ensured that his cartographic endeavour would be remembered by naming a lunar crater after himself: Langrenus, 132 km in diameter.

A more accurate method was needed and it was decided that this could be provided by observing the time of a lunar eclipse at different locations. Lunar eclipses were predictable so astronomers were able to prepare and install their observing equipment in advance and thus hopefully provide a more accurate longitude measurement, weather permitting. One such measurement was planned at the newly founded Royal Greenwich Observatory by the astronomers Edmond Halley (1656–1742) of comet fame, Robert Hooke (1635–1703), John Flamsteed (1646–1719) and Sir Jonas Moore (1617–1679). All observed the lunar eclipse of 26 June 1675 simultaneously with Giovanni Domenico Cassini (1625–1712), the French Astronomer Royal, who was based at the Paris Observatory. By comparing their data the exact longitudes of London and Paris could be established. Each observer except Sir Jonas Moore has a lunar crater named in their honour.

This with other observations of lunar eclipses over the following few decades established Greenwich and Paris as the leading astronomy centres of the world. Eventually the longitude at Greenwich became the reference meridian, marking the point of zero longitude. This also led to the establishment of Greenwich Mean Time which changes by one hour with every fifteen degrees of longitude west or east of Greenwich.[4] The 'longitude problem' (as it was known in its day) was eventually solved not by a scientist but by a clockmaker, John Harrison (1693–1776), who invented an ingenious series of clocks mounted in such a way that they remained level irrespective of the ship's motion or tilt and were unaffected by temperature or humidity, meaning that they could keep accurate time even when aboard ship for months on end.

When the Moon passes between the Earth and the Sun it completely covers the brightly shining solar disc, creating a solar eclipse, and darkness (totality) occurs for around seven minutes. It is these seven minutes of totality that makes observing a solar eclipse so spectacular. When the Earth passes between the Sun and Moon the Earth casts its shadow onto the Moon and a lunar eclipse is seen. A lunar eclipse lasts for an hour and, because not all the sunlight is blocked by the Earth, the Moon's surface attains a dull coppery orange colour. Thus observing a lunar eclipse is less popular and, while astronomical records of solar eclipses date back many thousands of years, the first record of a lunar eclipse appears in the Chinese records of 1136 BC.[5] If a solar eclipse occurs at the Moon's perigee or apogee (see p. 52) there is either an annular eclipse, where the Moon doesn't quite cover the Sun and a ring of light or 'fire' is seen around the Moon, or a total eclipse when it is closest. Unique to our particular moment in galactic time is the current positions of the Moon and Sun. While the Sun is 400 times the width of the Moon, the Sun as viewed from the Earth is currently 400 times further away than the Moon, and thus they look exactly the same size in the sky. Therefore, when they are on the same ecliptic (path across the sky) and they cross each other's orbit the Moon completely blanks out the Sun. In the past, when the Moon was closer to the Earth

Christopher Columbus
quelling the agitated
natives of Jamaica
by predicting
a lunar eclipse on
1 March 1504.

and further from the Sun it would have completely covered the Sun (always producing a total eclipse), and in the future when the Moon's orbit is further away from Earth only an annular eclipse will be seen. Solar eclipses are not rarer than lunar eclipses, at roughly two per year. At any one location on Earth it is harder to see a solar eclipse owing to the narrow viewing angle. When the Moon moves in front of the Sun, the Moon's shadow cast onto the Earth's surface is much smaller than the Earth, at only about 480 km across, creating a narrow viewing corridor. Outside this corridor only a partial eclipse can be seen. In contrast a lunar eclipse can be seen from wherever the Moon is above the horizon, which is over half the Earth.

The predicted or unpredicted occurrence of a solar eclipse has altered history on many occasions; there is only one recorded occasion when a lunar eclipse did so. This was in 413 BC, when the Peloponnesian War, a particularly savage war, was raging and the two Greek states of Sparta and Athens were fighting for supremacy in the Mediterranean. In Sicily the Athenian army was losing, and its commander, Nicias, a superstitious man, decided to evacuate the island and leave it to the Syracusans and

Spartans. Just as the nocturnal evacuation was about to take place, there was a lunar eclipse. The Athenians saw this as a sign sent by the gods telling them that they should wait for a further 'thrice nine days', according to Thucydides in his *History of the Peloponnesian War* (431 BC). This delay resulted in their fleet being completely destroyed and their eventual total surrender.

The actual appearance of the Moon during a lunar eclipse depends upon the Earth's atmosphere. As well as the normal copper colour, the Moon can appear blood red during an eclipse. In 1895 the reflected image of the African continent could be seen from London.[6]

Lunar orbits

Aristarchos of Samos (310–230 BC) used Euclidean geometry to try to determine the Moon's distance from Earth by observing the angle of the Sun when the Moon was half illuminated. This was a difficult exercise and Aristarchos' estimates were quite inaccurate. He found the Sun to be only twenty times further from Earth than the Moon is, whereas in reality it is 400 times further away. A century later Hipparchos of Nicaea (190–120 BC) improved on this by measuring a solar eclipse at two different latitudes – at Alexandria, where the Sun was 80 per cent covered, and the Hellespont, where it was fully covered. This allowed the Moon's distance from the Earth to be calculated accurately as 60.5 times the Earth's radius.[7] Using this method, the distance to the Moon would be 380,000 km, remarkably close to the actual distance of 384,403 km.

Using the orbit of the Moon around the Earth as a marker of time is more fraught with difficulty than one would expect. The Moon does not follow a simple circular orbit but an elliptical tilted one. This is complicated further by the fact that the Earth is spinning on its own titled axis (23°) around the Sun. The interaction between the differing orbits and spin leads to five different lunar cycles – the tropical, draconic, sidereal, anomalistic and synodic months. Most of us are familiar with the synodic cycle, which is the time between each new Moon: the time the Moon

Ptolemy, accompanied
by Urania, Muse of
Astronomy, measures
the lunar altitude
using a quadrant.

takes to go through all its phases. The synodic or lunar month
lasts around 29.53 days. The actual time the Moon takes to
orbit the Earth, the sidereal month, is somewhat shorter at 27.3
days. There are 29.5 days between successive full Moons, not
27.3, because the Earth and Moon have moved relative to the
Sun since the start of the Moon's orbital cycle.[8] The sidereal
month is more important for astronomical observations than

the Earthly one. The draconic month and anomalistic month are periods that take into account that the Moon and Earth tilt and wobble as they spin while tracing their elliptical orbits. These periods are used to calculate the appearance of eclipses. The tropical month is the time the Moon takes to travel through 360 degrees of longitude and lasts 27.32 days.

As the Moon traces an elliptical orbit, the distance between the Moon and Earth changes by around 40,000 km, or 11 per cent of the distance between the centre of the Earth and centre of the Moon (384,403 km). The Moon at its furthest orbital point, or apogee, is 406,700 km away, while at its closest approach, perigee, it is 363,300 km away. A further complication is that the Moon is gradually losing its spin and therefore spiralling away from the Earth by roughly 3.8 cm a year, equivalent to 1 km every 2.6 thousand years.

It takes the Moon one month to completely rotate on its own axis, which coincides with the time it takes to complete its

Louis Zimmer's astronomical clock of 1930, which includes the phases of the Moon, hangs from the Zimmer Tower, part of the city fortifications at Lier in Belgium.

The Moon being used
as a navigation aid.

orbit of the Earth. This phase-locking ensures that we always see
only one face of the Moon. The amount seen varies with the
Moon's orbital position. During its elliptical orbit its speed varies
so that an observer on Earth will be able to see it from different
perspectives, sometimes seeing a bit more around or behind the
normal western edge while seeing less of the normal eastern
edge, and sometimes the opposite, with more of the eastern edge
on show. This side to side view or optical libration of the Moon,
is also accompanied by north and south pole libration, this time
resulting from a 6°41′ wobble or deviation from the vertical. In
addition to a longitudinal and latitudinal libration there is a
third, so-called parallactic, libration which results from the daily
rotation of the Earth. This allows us to see an extra 9 per cent
of the lunar surface, with only 41 per cent of the total lunar
surface remaining hidden. The period between longitudinal
librations gives us the anomalistic month and between latitudinal librations the draconic month. Thus a peripheral feature like
a crater will seem to move around. Libration can be seen with
the unaided eye, but it has only been commented upon since the

invention of the telescope. The first to note these anomalies was the English mathematician Thomas Harriot (1560–1621), who in 1611 noticed that two peripheral features, Mare Frigoris and Sinus Roris, had moved between two observations three months apart.[9] It was left to other observers, notably Michael Florent van Langren and Galileo, to document this phenomenon in detail, the latter in his *Dialogue* of 1630. History had to wait for the genius of Isaac Newton (1642–1727) and his work on planetary motion before the cause of the phenomenon was explained. Interestingly Hevelius also described libration in *Selenographia*. Three of the maps in the book show the complete lunar disc, with the extra areas of the Moon that can be seen during libration.

Lunar libration is illustrated in these Hevelii illustrations in the elliptical sections added to the Moon's circumference.

Moon phases

The phase locking of the Moon's rotation gives the illusion on the Earth that it hangs stationary in the sky with the same hemisphere directed towards us. You can reproduce this phenomenon by placing a coffee table in the centre of the room (the Earth) and walking around the table, always facing it. While you, the Moon, orbit the table, you will be facing the table all the time but as you move around you will also see all views of the room behind the table (north, east, west and south), and your back will always face away from the table. Thus in the same way we do not see the other side of the Moon, sometimes known as the 'dark side'. It is not actually dark as it gets as much sunlight as the near side, and it is more fitting to call it the far side. We have known what the far side looks since the Soviets photographed it using their Luna probe in the 1950s. Until then there was much speculation, with views ranging from the sensible – that it resembled the near side – to the fantastic: that it would be Earth-like, with its own atmosphere, seas, life forms and populated cities.[10] Strangely enough, the far side is different to the near side; as it is more mountainous, has fewer Maria and displays more craters. This difference may be due to less volcanic activity rather than interplanetary impacts. We don't even know if the far side was always the far side.[11]

The new and full Moon are well known but there are actually eight recognized phases of the Moon which follow each other with a periodicity of 3.69 days: new, waxing crescent, first quarter, waxing gibbous, full, waning gibbous, last quarter and waning crescent.[12] Depending on the viewer's opinion the Moon glows white, silver or pale yellow. This glow is not

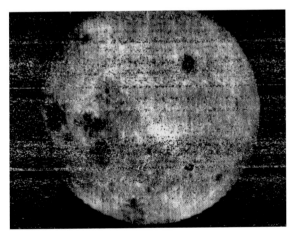

The first image of the far side of the Moon, taken by the Soviet Luna 3 spacecraft in 1959. The new features observed were given names such as Mare Moscoviense and Tsiolkovskiy crater.

due to its own light but that reflected from the Sun and Earth. The surface of the Moon is actually very dark, unlike Mars, which appears red because of its iron-rich soil. The distribution of light across the Moon's face gives us the different phases of the Moon, that is, the portion of the Moon that is on display, which can vary from 59 (due to libration) to 0 per cent. When the Moon is between the Earth and Sun, the side facing us is not lit by the Sun so is invisible in the night sky. This phase denotes the beginning of the lunar cycle, and has been known on Earth by many civilizations since time immemorial as the new Moon, marking a time of fresh hope or rebirth. The new Moon gradually moves from this alignment in the Earth's shadow and reflects progressively more sunlight. At first a small semicircle of light, or a crescent, appears on the east side of the Moon. The sharp line defining the phase (the boundary between light and dark) is the circular shadow of the Earth, known as the terminator. The now waxing (from the old English term for growing) Moon appears more prominently each night, and after a week the first quarter is reached and the complete eastern half of the Moon is illuminated. Here the quarter refers to the lunar cycle, not the amount of illumination. Over the next week or so the remaining half gradually fills, giving us the waxing gibbous Moon (derived from the old English word for hunchbacked, or humped). Therefore after around two weeks the whole lunar disc is fully illuminated, marking the full Moon. This light is directly reflected sunlight and also some secondary sunlight reflected from the Earth. Analysis of the reflected light from the Earth's surface by scientists at the European Southern Observatory in Chile has shown that the Earth is indeed inhabited by life.[13] Although this fact is not very revelatory in itself, it provides a novel method for looking for extraterrestrial life on planets orbiting other stars. If any of these so-called exo-planets have their own moons, then the reflected stellar light could provide observers on Earth the opportunity to look for the signature of life. The Moon reflects only 7 per cent of this light, around 0.2 lux, which is 400,000 times less bright than the Sun, but still some 100,000 times brighter than the stars.[14] At full Moon, particularly when it is at perigee,

A waxing gibbous
Moon observed on
a clear winter's day.

when there is snow on the ground and when it is a cloudless night, the moonlight can be strong enough to produce shadows – moon-shadows. One can easily walk around unaided in moonlight, but the light is not bright enough to activate the colour-sensing cells, or cones, in our eyes, so we see using only our rod cells (which sense brightness), and so moonlight appears as a cold silvery light. The moonlit landscape appears bathed in a mystical silvery sheen. To some this light is romantic, to others ghostly, but, either way, moonlight and moonshadow have a strong cultural influence on populations the world over. Nocturnal animals like the domestic cat prefer mating and hunting at the full Moon. On a wet night moonlight can even produce lunar rainbows, or moonbows. To our eyes moonbows seem white, but if photographed they resemble their colourful daylight equivalents. The reflected light also allows us to see the Moon during the day, as long as the Sun is not close enough in the sky to outshine the reflected light.

After the full Moon, the lunar surface begins to be illuminated only from the other side of the Earth and so the waning gibbous Moon is illuminated mostly on the western side. After a week the phase reaches the third or last quarter and here the opposite western half of the Moon is on display. The waning crescent appears after this, facing to the right (as the observer sees it). One lunar month later and the Moon is passing between the Sun and Earth and back to the invisible new Moon.

The waxing and waning is summed up neatly by Christina Rossetti's rhyme 'O Lady Moon, How to tell her age':

O Lady Moon, your horns point towards the east –
    Shine, be increased;
O Lady Moon, your horns point towards the west –
    Wane, be at rest.[15]

One of the mysteries of physics is that any spinning mass such as the Moon, Earth or Sun will create its own gravity. The greater the mass, the stronger the gravity. As Isaac Newton (1642–1727) discovered, while the force of gravity attracts other masses, the force of attraction decays quickly with distance. Thus the closer the masses are together the greater the influence. Consequently, for the Moon, Earth and Sun there is a triplicity of effects. To begin with, as the Earth rotates during the day, the Moon's gravity, although fairly weak, causes the Earth to bulge towards it by 2–3 cm. This does not have any noticeable effect on land but it does have a profound effect on the oceans, creating tides. As the Earth passes under the Moon's gravitational pull, any large mass of water is pulled towards the Moon. This bulge appears on the side facing the Moon and the side facing away. As the Earth rotates this bulge stays below the Moon, creating a high tide. The water perpendicular to the direct lunar pull must by contrast fall, so following the bulge sea levels drop, causing a low tide. Thus every twelve hours or so the sea level changes, so as the two bulges circle the Earth there is a high tide every 12 hours and 25 minutes. The changing orbit of the Moon advances the timing of each high tide by around 50 minutes per day. The Sun's gravitational field also plays its part, but as it is so far away its influence is weaker than that of the Moon. However, when the Moon and Sun are aligned, which occurs twice a month (at a new and full Moon), their combined pull will create an extra-large tide, a spring tide. When the opposite occurs and the Moon and Sun are at right angles to one another we get an extra low tide, or a neap tide, during the quarter Moon. These tidal terms are derived from old northern German or Dutch. The word 'tide' refers to a period of time. To the early coastal Europeans the ebb and flow of the sea must have had a profound influence on their lives. It would have dictated the times they could set sail and fish. The word 'spring' derives from the verb *springen*, not the season, and 'neap' derives from the old German word for minima.

Many important historical events have been pivotal on the timing of the tides, especially in the days of the sailing ship, an example being the English Navy's pursuit of the Spanish Armada in 1588 and its eventual destruction using fireships sent in on the tide. At lunar perigee the spring tide is at its greatest, and it has been suggested that it contributed to the sinking of the *Titanic* in 1912. A few months before the *Titanic* hit an iceberg, an exceptionally high spring tide had allowed extra-large icebergs to leave Greenland. Normally such a large iceberg would not be a hazard, as it would have melted before it could float south and reach the latitude of the famous collision.[16]

The orbiting Moon and Earth are each influenced by each other's masses, which interact to stabilize one another. The Earth without the Moon would spin more chaotically, leading to an ever-fluctuating climate of extremes – ice ages and warm periods.[17] Thus without the Moon's stabilizing effect life would not have evolved on Earth. If we go back many millions of years to the primitive Earth, when the Moon's orbit was closer and provided a stronger gravitational pull, the tidal ranges would have been huge. This frequent cyclical flooding and drying of the continental edges, particularly the warmer climates, provided the perfect biochemical conditions for life to gradually evolve from its completely aquatic environment to a littoral one. The tidal animals and plants we see today are left over from this time. The flora and fauna then colonized the non-tidal land and spread to cover the world, relying on rain and minerals to provide the biochemical medium essential to all life – salty water. Today the slowing of the Earth's axial spin and the recession of the Moon's orbit have less of a profound effect on biology, but still have a profound effect on human history and the economy.

If we look back even further, billions of years ago the Earth spun at such a rate that the day was only fourteen hours long.[18] The Moon would have been closer and had a stronger gravitational effect on the Earth, so that the tides would have been much higher and more energetic, occurring every seven hours. Over time the gravitational pull of the Moon (and Sun) on the oceans and the drag this generates robs the Earth of its angular momentum or rate of spin, making the day longer. As a consequence the Moon is slowly spiralling away from the Earth. We can even measure this rate of recession, thanks to the Apollo astronauts, who left Earth-facing reflectors on the lunar surface. By shining lasers at these mirrors and timing how long the reflected light takes to travel to the Moon and back, we have discovered that the Moon drifts further away by 3.8 cm per year. Thus as we lose spin our days will get gradually longer, by three to four hours in a billion years. The Moon, being further away, will then take longer to orbit the Earth. Thus 50 billion years from now, when the Moon is 540,000 km from the Earth, a month will take around 47 of our current days.[19] At this point the Moon's influence on the tides will be lost. The Earth's angular momentum will continue to reduce, though at a slower rate, owing to the solar gravitational influence on the tides.

Eventually the Earth's rotation will become slower than the Moon's and the lunar-induced tides will reappear. This time, however, the Moon will lose its angular momentum to the Earth instead and will begin to spiral inwards back towards Earth. Then, dramatically, at a critical distance of 18,500 km, the Earth's gravitational pull will be sufficiently great to tear the Moon apart, and Earth, like the outer planets such as Saturn, will have a ring of debris circulating the equator. However, this demise of the Moon is so far in the future that it is more likely that other catastrophic cosmic events will occur first, such as the Sun becoming a red giant and destroying both the Earth and the Moon.[20]

The Moon appears the same size all over the world, but while everyone experiences a full Moon, the exact view depends on your longitude and latitude.[21] The Earth spins at 15° per hour, which is faster than the Moon, so as the Earth spins the Moon rises in the east and moves across the sky to set in the west. If the Earth were motionless you would see the Moon do the opposite: rise in the west and over two weeks slowly move across the sky to set in the east. The Sun also crosses the sky east to west, but just to complicate matters, appears to move faster at 12.2° per day, therefore getting ahead of the Moon. At new Moon the Moon and Sun start in the same position in the sky and the Moon gradually lags behind, giving us the lunar phases. The Moon therefore rises around 50 minutes later each day, and high tide time progresses likewise, appearing every 12 hours and 25 minutes. The exact times vary according to location and the alignment of the Moon, Earth and Sun.

This waxing gibbous Moon glows orange because of smoke from wild forest fires.

One of a pair
of 'moon dogs'
or paraselenae
captured at dusk.

Atmospheric conditions alter the colour of the Moon, which can range from orange to blue. A blue-hued Moon has been reported at various times in history. These periods are associated with large forest fires or volcanic activity in which ash is thrown into the upper atmosphere where it can circulate for many months. The Krakatoan volcanic eruption in 1883 caused blue Moons for two years. On a cold night moonlight passing through ice crystals in the atmosphere can create lunar rainbows (different to moonbows) and, when slightly cloudy, halos, coronae and, most bizarre of all, 'moon dogs' or paraselenae, which are two areas of glowing light on either side of the Moon. The phrase 'once in a blue Moon' is thought to refer to the rarity of atmospheric blue Moons. A blue Moon can also refer to the second Moon in a month with two full Moons (although there is no difference in colour). The 'once in a blue Moon' phrase is used to denote a rare occasion, but in fact a double full Moon occurs on average every two years, plus or minus eight or nine months. In 2012 there were double Moons in March and August. These were followed by two new Moons in January and March 2014. These events are less noticeable, and are called black Moons. A black Moon has associations with the occult and witchcraft. The current metonic cycle (a nineteen-year period of lunar cycles) ends on

New Year's Eve 2028, so we will be able to see a blue Moon combined with a total lunar eclipse.

The view of the Moon does not just depend on astronomical mechanics and physics; psychology also plays its part. If you see the Moon rising over the horizon, particularly in the early evening when the Sun is low in the west, the rising Moon will look much bigger than normal. The effect is particularly spectacular if the moonrise is over buildings. This extra-large Moon is an optical illusion. It is easy to prove this illusion: just film the event and you will see the effect disappears. Philosophers have been debating the cause of the enlarged view for hundreds of years.[22] A recent study has shown that this illusion results from our stereoscopic vision and the way our brain perceives the size of objects at low angles.[23]

While we have inherited a Roman nomenclature for the months of the year, many cultures did not name the months

The Moon looks larger, especially when close to the horizon and distorted by atmospheric refraction, as in this view of Florence from Blount's *The Beauty of the Heavens* (1845).

directly, but named each full Moon instead. This is most evident among the North American Indians, with each tribe having its own specific names for each lunar cycle. The names usually refer to local environmental and seasonal conditions and are also influenced by geographical location. The winter Moons of December, January and February have names associated with cold and snow and the further north the more frigid the names: Cold Moon, Hard Moon, Severe Moon and Snow Moon are some of the names given by the northern tribes, whereas the Zuni of New Mexico called the January Moon 'ik'ohbu yachunne', which translates as 'sun has travelled home to rest'. With spring's approach, earlier in the south than the north, the names reflect the appearance of green trees, fish spawning and the return of migratory birds – Planting Moon, Flower Moon, Spring Moon and Noisy Goose Moon are examples that accompany the vernal equinox. By the summer the names celebrate the abundance of flowers and fruit, for example, Flower Moon and Strawberry Moon. Autumn names reflect the ripening of corn, nuts and fruit – Nut Moon, Ripe Corn Moon and Salmon Moon. The September Moon was often known in North America and Europe as the Harvest Moon as, close to the autumn equinox, the three nights of extra light provided by the full Moon were used to finish gathering in the remaining crops. The October Moon is widely known as the Hunter's Moon or Blood Moon, reflecting the fact that this is a time for harvest or killing surplus game like bison and deer before the winter hardens.

In present-day China the names are more esoteric, with April being the Peony Moon, June the Lotus Moon and September the Chrysanthemum Moon. Here August is named as the Harvest Moon.

The time we use today originally derived from observations of our surroundings, following the rising and setting of the Sun, giving a sense of day and night, while the nocturnal procession of the Moon's phases allowed us to mark the progress of the seasons. These easily observable celestial objects crossing the sky according to the physical laws of motion provided early humans with primitive ways of keeping time. Marking the passage of full

An Inuit moon mask.

An Inuit depiction of an anthropomorphic moon with a leashed polar bear and a bucket.

64

and setting Moons allowed them to plan ahead and to annotate their past. By dividing the lunar month into quarters, we arrived at weeks with seven days, with names inspired by the celestial bodies defining their existence, such as Sunday and Monday. As timekeeping became more sophisticated the anomalies between the lunar and solar cycles appeared. Alongside these anomalies, the quest to synchronize and standardize calendars has led to ever more precise definitions of the orbital times of the Earth, Moon and Sun. Thus our sense of innate time driven by our natural surroundings has gradually moved away from the Moon and Sun, and now we follow atomic time, a time derived from the precise and infinitesimally small period it takes electrons in a caesium atom to flip between two energy levels. We have effectively moved from the very big to the very small to determine time. At the very small, or quantum, level change is rare whereas at the very big level it is common. On a large scale gravity can alter the orbits of the Sun, Earth and Moon, alter our tides and drive evolution. By using atomic time with precision to the nearest trillionth of a second, we can measure distance with equal accuracy and

The Chinese Peony Moon.

measure how far away the Moon is from Earth to the nearest centimetre. While we have observed the Moon for many centuries and catalogued its effect on our physical environment, it has been even more influential on our beliefs, as the next chapter illustrates.

3 Stories and Legends

Virtually all of the world's cultures and ancient civilizations, including the Persians, Greeks, Romans and Chinese, have their own stories and legends about the Moon – as a god, as a messenger, or as symbol of good or evil. To some cultures the Moon is male, to others female; to some a symbol of death, to others rebirth. The awe and sense of mysticism that early humans felt towards the waxing and waning Moon connect to our primal fascination with the Moon today. Throughout history much architecture has been inspired by the Moon and the beliefs associated with it, such as Neolithic stone circles, the Aztec Pyramid of the Moon in Peru, and the Monastery of the Moon in Lebanon. The Moon is found in much religious symbolism. It is used in images of the Crucifixion and the crescent Moon is a symbol of Islam.

The two most visible objects in the sky are the Sun and Moon, and when viewed with the naked eye they look like flat discs rather than spheres. They both shine, one providing a warm, yellow light, the other a cold, silver light. It is easy to see why to early man the Sun and Moon were supernatural objects, the Sun symbolic of life, a reminder of the daily and yearly cycle of life, while the Moon, constantly waxing and waning, represented birth and death. These two highly visible celestial objects were worshipped and feared and virtually every culture has striven to understand their role in the heavens.

To the ancient Greeks, Artemis or Selene was the name given to the Moon, a goddess. According to mythology Selene had 50 daughters (representing the 50 Moons between each Olympian cycle) by Endymion, her male companion. Endymion, who was a mortal (either a shepherd or of royal blood – the myth varies) was caught in eternal sleep, a state granted either by Zeus or induced by Selene herself, resulting in Endymion's retaining his youthful beauty. Selene is said to have visited him in his cave on Mount Latmus every night, just as the Moon visits the sky every night.[1] In the medieval version of this myth, Endymion becomes immortalized as the Man in the Moon.

In other cultures the Sun and Moon were given human characteristics and behaviours. Often they were seen as brother and sister, husband and wife, male and female, but never as twins. To the Australian Aborigines the Sun is female and the Moon, male. These views are reflected in our modern languages: in English, French, Italian, Latin and Greek the Moon is feminine, while in the Teutonic languages of Northern Europe, such as German, it is masculine.

G. B. Cima da Conegliano, *The Sleep of Endymion*, c. 1505.

Some argue that Moon worship, or luniolatry, is older than that of the Sun. Assyrian monuments show bas-reliefs of the Moon as a sacred symbol. The male Moon god Sin, 'the Lord of the Moon', was considered the most important deity long before the Greeks later promoted the Sun to this position.[2] We derive the name Sinia from this origin and Mount Sinai could have once been dedicated to the Moon. In old Hebrew the Moon was called the 'queen or princess of heaven', and the Jews may have shown some religious respect for the Moon from the times they inhabited Egypt.[3] As with other ancient civilizations, the ancient Egyptians attached great import to the Moon. The Moon was not considered a deity in itself but represented manifestations of specific deities. The full and crescent Moons are represented in many Egyptian myths, associated with gods such as Thoth, Osiris and Horus, and is symbolically represented as a disc or an eye. The Moon also serves as a metaphor for both youthfulness and death, depending on each lunar phase.

In Sanskrit the waxing and waning phases are known as 'Shukla paksh' and 'Krishna paksh', white and dark side respectively. The two periods are further divided into sixteen parts, beginning with the black phase and finishing with the white phase. To the Buddhists these link the negative and positive

An ancient Greek or Roman gold ring depicting Selene sitting on a crescent Moon and surrounded by stars.

An ancient Egyptian turquoise ring from the time of Amenhotep III, showing a baboon and feather under a full and a crescent Moon.

Chandra, the Moon,
represented as the
god Vishnu.

aspects of life; when the lunar month is quartered, each part is believed to correspond to one of the four human states: awake, dreaming, sleep and deep sleep.

The pagan Moon

When Christianity swept across Europe and the Roman Empire, the Church authorities and monarchs needed to replace the outdated pagan beliefs with those of the Bible, and many pagan festivals were replaced by the Christian ones we know today. The Moon held an important place in paganism as stated by King Canute (990–1035) in his set of laws: 'Paganism is when one worships devil-idols . . . one worships heathen gods, and Sun or Moon, fire or running water.'[4] Centuries after the Romans had left Britain, the idea of paganism was still seen as cause for concern. The bishop of Noyon (588–660), later St Eligius, left us in one of his sermons a long list of behaviours considered pagan that should consequently be avoided by good Christians. Some of these involved the Moon, and included: 'Do not wait for a particular day or phase of the Moon to begin something'; 'Do not shout at the eclipse of the Moon'; 'Do not call the Sun or Moon "masters" or swear by them.'[5]

In 58 BC Julius Caesar, writing on the Gallic wars in his book *Commentarii de bello Gallico*, described his perplexity at the Suebian King Ariovistus' refusal to re-engage the Romans in further battle after fighting all day. He eventually found out from captured prisoners that it was due to a custom in which the chief's wives would, through divination, pronounce on the outcome of a battle. In this case there was a full Moon under which, the wives proclaimed, any battle fought would be lost.

Although paganism was eventually forgotten, the Moon remained important to Christian iconography. In the book of Revelation, the Virgin Mary is described as 'a woman clothed with the Sun, and having the Moon under her feet, and upon her head a crown of twelve stars'. Catholic painters and sculptors have used this imagery in many churches and religious art. Two good examples can be found in southern Germany. The first is in Marienplatz in the centre of Munich where, atop a column in the centre of the square outside the Rathaus, the Virgin Mary cradles an infant Christ and stands proudly on a gilded crescent Moon. The second is inside the Cathedral (Dom) of Trier, where a statue depicts the Virgin Mary and Christ, with Mary standing on a crescent Moon, this time ungilded. The crescent has a face depicting the man in the Moon.

Another Christian symbol linked with the Moon is the hot cross bun, the spiced fruit bun consumed at Easter time. The origins of hot cross buns date back to pre-Christian times. They

In the Marienplatz at Munich, a gilded Virgin Mary holding the infant Jesus stands on an inverted crescent Moon.

were consumed in ancient Greece and have also been found in the Roman ruins of Herculaneum in southern Italy. In this era the buns were eaten all year round, and the cross honoured Diana, the Roman moon goddess. By post-pagan time this had changed, with the round bun representing the full Moon, and the cross marked on top representing its four quarters. It was not until Tudor times, when Elizabeth I declared that spiced buns could only be sold at Easter and for funerals, that the hot cross bun became associated with the Christian festival of Easter and the Crucifixion. This association served to strengthen the link between Easter and the full Moon.

The Moon, particularly in crescent form, is closely associated with Islam. From ancient times the Moon was of religious importance, not as an object of worship but as an important tool for calculating important religious festivals such as Ramadan. The adoption of the crescent Moon is believed to have begun when the Turks conquered Constantinople in 1453. Before the city fell the crescent Moon was displayed ubiquitously, representing the moon goddess Diana, the city's patron. The crescent Moon was adapted to symbolize this great Ottoman victory. Eventually, however, it became a symbol of the Ottoman Empire and Islam. Today a white crescent Moon is still part of the Turkish flag, where it is accompanied by a five-pointed white star on a red

A plateful of hot cross buns girdled with Easter daffodils.

background. The flag is called 'ay yildiz', or moon star. The star is thought to represent the morning star, as described in the Koran. It could equally represent Venus or Mercury, as these are often seen in the same region of the sky as the Moon. Over the centuries many Islamic nations and organizations have adopted the crescent moon and star as a religious symbol. The Islamic version of the Red Cross is represented by a red crescent moon on a white background. The design is repeated in at least ten other national flags: Pakistan and Singapore have a white crescent moon and star, while a red crescent moon and star are found on the Algerian and Tunisian flags; the Malaysian flag sports a yellow crescent moon with a fourteen-pointed star, representing the country's fourteen states. Other national flags with lunar crescents and stars include those of Mongolia, Nepal, the Maldives, Mauritania, Turkmenistan and Uzbekistan. The crescent moon can be found on the top of minarets on mosques around the world.

The flags of the Maldives and Mauritania, visual examples of the Islamic link with the crescent Moon.

The Moon in folklore

In folklore the Moon is often associated with the tide, water, moisture and dampness. For example, in medieval Europe those who made their living from the sea believed that the cycle of life and death depended on the tide, with death occurring more commonly during the falling tide. Indeed, 'The Moone is found by plain experience, to bear her greatest stroke pon the seas, likewise in all things that are moist, and by consequence in the brains of man.'[6] The Persians thought the Moon (female) was the cause

Luna, high above a
16th-century northern
European landscape,
riding the heavens in
her chariot, engraved
by J. Sadeler after
Martin de Vos.

An engraving by
N. Dorigny after
Raphael of Diana as
Moon goddess (then
thought to be Earth's
closest planet) in the
House of Cancer.

of an abundant water supply, rain, warmth, wealth and riches. Some of the most fruitful and verdant places in Persia were compounded with the word 'Mâh', meaning Moon. The Moon was thought to enhance plant growth, a belief also held by the Egyptians.

In pre-industrial Europe the Moon was associated with dampness and its appearance was believed to predict the weather. A new Moon on a Monday, for instance, was seen as a sign of good weather ahead.[7] Folklore held that the quartered Moon could hold water and therefore influence rainfall. It was posited that the turned-up arms of the waning crescent Moon acted as a basin and could collect water, creating dry weather, while the waxing crescent Moon points to Earth, causing the water to be lost and creating rainfall and damp weather. There is little factual evidence that there is any connection between rainfall and lunar phases.

In astrology, the Moon rules Cancer, a water sign, and is feminine and receptive. The Moon's feminine qualities link her with women, and her gravitational force influences women's monthly cycles as well as Mother Earth's water realm through its effect on tides. The mysterious Moon only reveals half of herself to us. Nevertheless her appearance changes constantly because of her phases.

There is a tarot card representing the Moon. The depiction differs from set to set but the majority show the Moon as both full and quartered with two pillars, one on either side. The mid-ground shows a wolf and domestic dog howling at the Moon, while in the foreground a crayfish or lobster emerges from a pool of water. The consensus on the symbolism of these objects varies among tarot readers, but most suggest that the Moon symbolizes confusion and mistrust.

A European wolf baying at the Moon, a behaviour often associated with werewolves.

In Tudor times divination and astrology were common practices, and the casting of horoscopes was taken very seriously. In a letter to her grandmother, Arbella Stuart (1575–1615), a claimant to Elizabeth I's throne, mentions that she has enclosed her hair, which was cut on the 'sixth day of the Moon'. This

lunar precision was necessary as the hair was going to be used astrologically to cast a horoscope to foretell Arbella's chances of gaining the English crown.[8]

In Victorian times, in some parts of northern England, it was considered a sin to point at the Moon, while gentlemen would touch their hats and young girls curtsy to the new Moon.[9] There were many other Moon-related superstitions: the new Moon shining on your purse would keep you poor; only make wine in the dark of the Moon; calves weaned during a full Moon produce the best milking cows; sleeping in the moonlight can make you blind. To this end many thatched cottages have extra thatch extensions projecting over the upstairs windows that are designed specifically to prevent moonshine from falling on the sleeping inhabitants. Excessive exposure to moonshine while sleeping was also thought to cause birth defects and alter one's mental health, and was associated with diabolical possession.[10]

This diabolic connection is demonstrated by the cultural belief in the werewolf, according to which exposure to the full Moon causes the tortured metamorphosis of man into beast. Werewolves are first found in Greek mythology. Ovid (43 BC–AD 17) wrote in the first volume of his *Metamorphoses* a story about King Lycaon, who illicitly served human flesh to Zeus, who then

A non-scary Halloween witch flying on a broomstick, with a crescent Man in the Moon in the distance.

punished the king by turning him into a werewolf. Thus from Ovid and Lycaon we get 'lycanthrope', a term for werewolf. The legend of lycanthropy or werewolves spans many cultures, particularly in northern Europe. Some nineteenth-century psychiatrists defined it as a form of madness where someone believes themselves to be an animal such as a wolf, in extreme cases practising cannibalism, or a more spiritualistic interpretation where a physical transformation into a wolf-like animal takes place, again engendering a thirst for flesh, including human. The werewolf has become ingrained in our culture in this latter form. In either case the lycanthrope is mad and usually performs his or her gruesome deeds during a full Moon, or sometimes the new Moon.[11] Nowadays the werewolf only exists in fiction, particularly of the horror and fantasy genre, in films such as *An American Werewolf in London* (1981), in books such as the Harry Potter series and the Twilight series, and Michael Jackson's 'Thriller' video.

Another diabolic association with the Moon is that of witches flying on broomsticks at full Moon and therefore able to control the Moon. In Shakespeare's *Tempest* (Act v), Prospero says: 'His mother was a witch, and one so strong / That could control the Moon'. In Eastern Europe before the twentieth century, mermaids representing drowned maidens were believed to dance under the Moon in order to attract young men to the water before drowning them.

Moon gardening

Even today there is a widely held belief that the Moon influences plant growth. This belief has influenced folklore and horticulture for centuries. Ancient scholars such as Plutarch (AD 45–120) and Pliny (AD 21–79) considered the Moon to have a positive influence on the growth of flora and fauna. The mechanism behind this lunar phenomenon is believed to be its gravitational tidal pull on the ground water, with a raised water table increasing the moisture content of the soil and encouraging germination and plant growth. The extra moonlight as the full Moon waxes and wanes could also contribute to growth, but this would vary with cloud cover. With this reasoning in mind, it is thought best to plant during the waxing Moon and weed during the waning Moon.[12] Old English folklore has more to say on the matter: for example, that vegetables such as turnips and leeks grow better during a waxing and full Moon. The potency of a medicinal plant is thought to be greater if it is picked during the full Moon. This belief even extended to cutting timber. In the first century AD Pliny stated that trees should only be felled during the waning Moon, when a drier, better-quality timber would be produced the waxing Moon would cause the sap to rise. These beliefs were incorporated into eighteenth-century law in France, where tree-felling was only legal during the appropriate lunar phase.[13]

This lunar influence was treated with scientific reverence in the sixteenth century. During this period formal gardening on a grand scale began in the UK, with designs copied from those already established on the Continent, particularly Italy. The botanist John Gerard (1545–1612), famous for publishing *Herbal* in 1597, and owner of the first botanical garden in the UK, believed the lunar phase was important when planning a garden. Gerard was employed to maintain the gardens at Theobalds Palace, Hertfordshire, the grand house of William Cecil (1520–1598), the wealthy and long-serving secretary of state and Lord High Treasurer of Elizabethan England. This was reported to be one of the most exquisite gardens of its time, designed to impress Queen Elizabeth I. The Moon played an important part in the garden's planting calendar. Beginning in the early spring, vegetables would be planted only during a waning Moon. Then with each subsequent new Moon flower seeds would be sown, such as marjoram, violets, rosemary and lavender. Seedlings such as the marigold were planted out during the waxing Moon, while pansies (the queen's favourite) were planted out only during a waning Moon. Gerard believed that this provided the best results, producing bigger, broader and more colourful flowers. Anything that Gerard planted during the waxing Moon had its flowers harvested during the waning Moon, especially when the skies were clear. Fruit was gathered at the same time, as it

Sir Walter Raleigh, 1588. A crescent Moon over a patch of wavy blue water alludes to Elizabeth I as the moon goddess Cynthia. His costume is covered in pearls, symbolizing Elizabeth as the Virgin Queen.

was thought that fruit gathered during a waxing Moon would taste bitter.[14] Today similar lunar gardening is practised by many across the world.

In these Elizabethan gardens the Moon played a central symbolic role, as Queen Elizabeth the Virgin Queen was often portrayed as Diana or Cynthia, the Moon goddess (the goddess of chastity and hunting). To cite an example, in 1591 on her summer procession Queen Elizabeth visited the Earl of Hertford's estate in Elvetham, Hampshire. The earl was keen to impress her in order to improve his standing at Court. By the time the queen accepted the invitation the earl had only a few weeks to transform his grounds. He recruited an army of tradesmen and labourers who managed just in time to create a huge crescent

Moon-shaped lake. This lake provided the stage and backdrop for a nightly series of spectacular nautical pageants.[15]

The cult of Elizabeth as Diana or Cynthia became popular as her reign reached its zenith and was an important part of her public image. In this regard many of her official portraits from 1586 to her death in 1603 show symbolic representations of the crescent Moon, usually as a jewelled brooch added to her dress or worn in her hair. The famous and symbolic *Rainbow Portrait* of 1600–1602, one of the last commissioned paintings, shows Elizabeth elaborately dressed in a gown that was designed for a masque, wearing a half-crescent moon jewel in her hair.[16] Her private accounts from this time show that many of the gifts she received from her courtiers were gold bejewelled half and crescent Moons. Sir Walter Raleigh, who was the most prominent of these courtiers, is attributed with popularizing the Cynthia cult. He even wrote an epic poem entitled 'Book of the Ocean to Cynthia', which was never finished as Elizabeth died during its composition. In one of the principal portraits of Sir Walter, dressed in the queen's colours of black and white, a small crescent Moon is clearly visible in the top left-hand corner. The crescent Moon represented not only chastity and virginity but, to people like Raleigh, England's dominion over foreign lands. On a personal level the Moon symbolizes Raleigh's willingness to be controlled by the queen, just as the Moon controls the tides. This post-Armada period marked the beginning of the British Empire. Likewise, Henry II of France used the crescent Moon as a device to reflect French expansionism. Thus in Tudor England the Moon symbolized Elizabeth I as Gloriana, a symbol of purity and strength in the world both old and new.

The Man in the Moon

When we look at clouds our brains often make out familiar shapes, such as animal or human forms. This anthropomorphic synthesis and psychological phenomenon known as pareidolia is hard-wired into our brains. Thus as humans across time have stared up at the luminous and blotched lunar surface they have imagined human and animal forms in the patterns that mark the plains and mountain ranges of its near-side surface. In Europe we see a male face – the Man in the Moon. In other cultures it is a woman or a hare. In European culture the Man in the Moon is one of the strongest and most popular lunar associations. It is pointed out to every small child, along with the idea that the Moon is a world made of cheese. Of course to a child these ideas are not strange and often go unquestioned. This Victorian nursery rhyme of uncertain origin portrays this childish view:

> The Man in the Moon
> Came down too soon
> And asked the way to Norwich
> He went by the south
> And burnt his mouth
> With eating hot pease porridge.[17]

The Man in the Moon myth revisited in a print of 1880.

In folklore drunkenness is associated with the Man in the Moon. During a lunar eclipse the red face of the Man in the Moon

was supposedly a result of his drinking too much claret. Moonlight is another name for smuggled spirits (and perhaps a source for the word moonshine).[18] Moonlighting is a term used to describe taking on extra work at night, after the 'day job' has finished.

Medieval folklore provides the origins of this lunar phenomenon. In German folklore the Man in the Moon started as a poor man who defied convention by working on the Sabbath instead of attending church, so his punishment for collecting firewood and not resting was to be banished to the Moon, where he was condemned to stand for ever as a warning to all Sabbath breakers.[19] In other versions of this tale he builds a fence or steals coal from his neighbours. The German version became common across Europe and the Man in the Moon is often shown with a bundle of thorns on his back. A well-known image is in Gyffyn church in Conway, Wales. The first mention of this folklore in English was in 1157, when the abbot and scientist Alexander Neckan of St Albans (1157–1217) commented on the vulgar belief about 'a rustic in the Moon who carries the thorns'. In the English version the man is banished to the Moon for stealing a bundle of thorns. Dr Peacock (1393–1461), bishop of Chichester, commented in 1449 on 'this opinion that a man which stole a bundle of thorns was set in to the Moon, there for to abide forever'.[20]

To Shakespeare the Man in the Moon myth was a literary gift not to be missed. From *A Midsummer Night's Dream* (v, i) we have: 'All that I have to say is, to tell you that the lantern is the Moon; I, the man i' the Moon; this thorn-bush, my thorn bush; and this dog, my dog.' *The Tempest* (ii, ii) includes the lines: 'Caliban hast thou not dropped from heaven? Caliban I have

The Japanese associate hares with the Moon, as in this 19th-century woodblock print by Hiroshige.

The Man in the Moon with an attendant star and distant planets, a contemporary painted wood decoration from Indonesia.

seen thee in her, and I do adore thee: my mistress showed me thee and thy dog and thy bush.'

In some parts of the world different images are seen on the Moon's surface. The Chinese, Mongolians and Aztecs have various fables concerning a hare and how it ended up on the Moon. In one such fable a hungry Buddha comes across the hare, who kindly offers to help him. Buddha replies that he is hungry but has nothing to offer the hare. The hare offers himself as a meal, so Buddha lights a fire and the hare jumps in. The divine Buddha then grabs the hare from the flames and throws him into the air, setting him on the Moon, where he remains to this day. This association is strengthened by the fact that the gestation period of the hare is around 30 days, close to a lunar month.[21] A modern view from the USA is the basketball player in the Moon. The

player's head is described by the Mare Imbrium and his shooting
left arm by the Sinus Roris, both on the upper western quadrant.
The ball is seen on the upper right quadrant, marked by the
Mare Serenitatis. In some Pacific cultures the lunar markings were
once thought before modern times to be forests where groves
of exotic trees grew.

In 1895 the schoolteacher Miriam Levy was inspired to find
out from the 555 children in her school, aged between four and
fourteen, what they thought about the origins of the Man in the
Moon. The most popular answer to her question of how the Man
in the Moon got there was simply 'No idea'. Among the eight-
to nine-year-olds who offered an explanation, the most popular
reply was that he had got there by balloon or flown there, or that
God had put him there. Others of the same age answered that
he was born there, he ended up there when he died, he is an angel,
or simply by electricity. The older children did not think there was
a Man in the Moon, and gave mostly rational and scientific
explanations.[22]

The special Moon

Since ancient times the white reflected dish of the full Moon has
been likened to the metal silver. For instance, the Greeks made
shrines to Diana out of silver. To astrologers and alchemists the
physical properties of silver were manifestations of the Moon's
influence. The only purported way to kill a werewolf is to shoot
it with a silver bullet.

In Hinduism Chandra is a lunar deity, as well as lord of the
plants and vegetation. The waxing Moon is considered to be
beneficial, particularly the full Moon, while the waning and new
Moon is thought to be corrupting and maleficent. In Hindu
astrology the Moon is a planet and those born under it gain
wealth and happiness. Chandra is a common surname in India and
in the USA it is used as a girl's name.

In North America, where tribes have a long association with
the Moon and the seasons of the year, the Moon also influenced
funeral rituals. Following the death of a tribal chief, it was customary

for his widow to sit every day for 28 days (through a full lunar cycle) under a totemic tree bearing the dead chief's tomahawk and bow and arrows. To this end many totem poles have lunar symbols carved upon them.[23]

Meztli was the Moon deity of the Aztecs and 30 miles north of Mexico City are the remains of the ancient city of Teotihuacán, built around 200 BC. The ruins are full of huge stepped pyramids, the largest of which is dedicated to the Sun. Among the other pyramids is the so-called Pyramid of the Moon, dedicated to the Moon and aligned with the Avenue of the Dead. This pyramid arrangement is repeated throughout South America, and expanded by the Incas who often built Sun temples with smaller Moon temples attached.[24]

In twelfth-century Cambodia the Hindu temple of Angkor Wat was built by Suryavarman II. Its dimensions are influenced by Hindu observations of the lunar cycle. To the Hindus the Moon had 27 to 28 'lunar mansions' for each of its different paths through the sky.[25]

Even today lunar rituals are practised and can be dangerous. Deaths have occurred as recently as 2012, as reported last year by four Indian doctors in the *Australian Medical Journal*. This ritual starts before sunrise with the collection of bark from the *Alstonia scholaris* (or blackboard tree). During the day a concoction

The Pyramid of the Moon and its plaza at the Mesoamerican city of Teotihuacán, Mexico.

is made from the bark, ready to be drunk at the full Moon. Normally there is nothing dangerous about this ceremony, but sometimes in the early morning light the bark from the *Strychnos nux vomica* tree is mistaken for the similar-looking *Alstonia* bark. Unfortunately *Strychnos* bark contains the neurotoxin brucine, which can be lethal if ingested.[26]

The Moon's physical presence and influence on the human psyche and imagination is ubiquitous. Firmly embedded in the world's religions and mythology, the Moon has inspired human thought throughout history. With this close universal association between humankind and the Moon it does not require a great leap of the imagination to suspect that this link is not just some ethereal attraction but a hardwired biological one. After all, if the Moon influences the tides, then why over the millions of years of human evolution would we not have become sensitized to it in some way? This suspected biological influence on our physical and mental health has been documented since the ancient Greeks and is explored in the next chapter.

4 Man and the Moon

Lunation has been thought by many to guide human circadian rhythm and thus influence health, behaviour and mental state. Superstition underlies many of these be liefs, such as the idea that surgery or giving birth should be avoided during a full Moon. The Bible insinuates that overexposure to the full Moon is bad for one's health, leading to the observer becoming moonstruck, or a lunatic. The full Moon has for many years been thought to affect mental health, with reports of suicide rates increasing with the full Moon, and the raised frequency of epileptic seizures. In legend the full Moon is accredited with the transformation of man into the fearsome werewolf.

Today we run our lives according to a 24-hour clock, enmeshed within a seven-day week. Our calendar is no longer set by celestial time but regulated by nuclear clocks. However, we are still biologically wedded to a celestial clock synchronized to the solar cycle of day and night. It is the duration of each light and dark phase of this cycle that influences our behaviour and biology. Most of us spend the night asleep and inactive while during the day we are awake and active. It is on top of this basic diurnal variation, or so-called circadian rhythm, that we set our daily habits metered out by clock time. The circadian rhythm is orchestrated by a complex arrangement of organs and tissues within the brain, such as the pineal and pituitary glands. We are exquisitely sensitive to this daily light and dark cycle, as anyone who suffers from jet lag or stays up late will know. Long-term disruption of the circadian rhythms can lead to obesity, cancer and reduced life expectancy.[1]

We try to minimize these celestial influences on our lives: when the weather is cold or hot we heat or cool our rooms; when it is dark we switch on the lights. We spend ever-increasing amounts of time indoors both during work and leisure time, and even when we travel around it is in a controlled environment. Nature's rhythms are gradually becoming hidden from us. This is a recent historical phenomenon, as in pre-digital, pre-industrial times our day was dictated by the daily light and dark cycle.

Throughout history scholars have speculated about a link between the regular cycles of the Moon and our behaviour and well-being. Consequently there has been an enormous amount written on the subject and it takes a great deal of careful investigation to distinguish between myth, pure coincidence and fact. While we cannot yet completely disregard any subtle lunar influence on our physiology or behaviour, we can certainly dismiss the views of those who claim there is a direct lunar influence on mental health or child development.

While there are many examples of animals, particularly marine creatures, that seemingly alter their behaviour according to the lunar phase, the effects are secondary, and result from changes in the tide or brightness of the evening sky. The duration of these specific biorhythms varies and can be circaseptan (weekly) or circatrigintan (30-day or monthly).

In humans, one of the most obvious circatrigintan, or monthly, cycles is the menstrual cycle. The name derives from the Latin *menses*, for month. Moon, month and menstruation are therefore already linked etymologically. This complex cycle is driven by hormones, controlled by many factors and influenced by a woman's health, both physical and mental. The cycle time varies from woman to woman but is generally between three and five weeks, averaging around 28 days (the lunar synodic period). Is there a link, or is it just another lunar coincidence? Both Aristotle (384–322 BC) and Hippocrates (460–370 BC) thought there was a strong link and that menstruation began with the waning Moon. They linked this with the premise that sexual powers increased until the full Moon and decreased with the waning Moon. While common experience suggests that the duration is coincidental, as other mammals have cycles longer or shorter than the lunar cycle, could the Moon have some subtle effects specific to humans? Anecdotal evidence comes from some women who feel that they are mystically entrained with the lunar cycle. Of course, from a statistical viewpoint some women will have periods that occur close to the synodic lunar period. Studies testing whether this link is real or imaginary have largely shown no direct connection. Two studies from the 1980s involving a sample of 1,100 women suggested that the two cycles were synchronized, but these are the exceptions and nowadays there are no demonstrable links.[2] Those advocating a link between the lunar cycle and reproductive function raised the fact that pregnancy lasts 40 weeks or exactly ten lunar cycles, the significance of which holds no scientific merit today.

The duration of the lunar orbit changes with time and millions of years ago was shorter, so it could have been during the evolution of our early hominid ancestors that the length of the menstrual cycle was set. For the lunar cycle to influence the biological menstrual cycle there would have to be some biological advantage to the breeding female. If finding a mate was a nocturnal process, then a full Moon may have helped. Equally, ovulation occurring during a new Moon may support this notion. Hence fertility could have been improved by linking

Moonstruck women deserting their husbands in a 17th-century Dutch engraving.

the two cycles. Nowadays, particularly in the leading economies, controlled environments with artificial lighting and constant room temperature have uncoupled the daily and seasonal influences from our biological cycles. Thus if any vestigial links remain between our ability to procreate and the Moon, they are now redundant.

Mental health and lunacy

While the boundaries of astronomical knowledge progressed and expanded thanks to the great advances made by Greek and Arab scholars, mathematicians and astronomers, the view that the Moon influenced the human condition remained static. Galen, Aristotle and Hippocrates all thought that insanity waxed and waned with the Moon. They referred only to conditions that were intermittent and involved convulsions today known as epilepsy (a word derived from the Greek term for 'seize').[3] The notion that the Moon is a direct cause of madness gives us the etymological link to lunacy and lunatics. Lunatics were often considered to be 'moonstruck'. This link is also reflected in other languages: in French there is *avoir des lunes*; in Italian *lunatico*; and Latin *lunaticus*, all old terms used to refer to people with mental health conditions such as epilepsy.

The ancient Greeks believed that the Moon's influence was opposite to that of the Sun, whose properties were to dry, desiccate and generally promote water uptake. The Moon in contrast was associated with water, dampness and precipitation. Noticing that on cloudless nights the Moon and its associated moonshine seemed to impart a cool, moist and refreshing feel to the Mediterranean night air, and a copious formation of dew in the early morning (in reality because air temperatures are usually lower under a cloudless sky), the ancient Greeks presumed that the Moon also influenced the flux of body fluids, such as the blood and the cerebrospinal fluid surrounding the brain. Ironically the association between the Moon and tides was never made by the Greeks, probably because of the small tidal range of the Mediterranean.

Alchemical symbolism, a coloured image from Salomon Trismosin's *Splendor solis* (1532–5), the Moon above the queen represents the dissolving 'lac virginis' (mercury) while the Sun above the king is the coagulating masculine principle (sulphur).

An example of the inhumane treatment endured by people suffering from mental health problems in Sir Charles Bell's *The Anatomy and Philosophy of Expression as Connected with the Fine Arts* (1844).

In the medieval era, the progress of medicine slowed, and physicians made little effort to distinguish between astrology and astronomy. In their minds illness was due to an imbalance between the four humours – blood, phlegm, black bile and yellow bile – and these humours were in turn influenced by the planets (including the Moon, which was then considered a planet): blood was influenced by Jupiter, black bile by Venus, yellow bile by Mercury and phlegm by the Moon. The medieval physician would thus regard a phlegmatic person to be the most sensitive to lunar quartering. From an astrological viewpoint a phlegmatic person would be more reflective and sensitive.[4]

To the physician, the lunar phase was considered important. Treatment under a full Moon would be aided by the Moon's ability to rejuvenate and enhance survival. The chances of surviving an acute illness were gauged to be best when the Moon was full, while those unfortunate enough to fall ill during a waning Moon were considered the least likely to survive. When it came to surgery, barber surgeons (medical practitioners who practised surgery, unlike their colleagues who did not) were observant of lunar phases and would time their bloodletting sessions accordingly. While most surgeons and physicians endorsed these beliefs and superstitions, some physicians and scholars began to systematically question them.

One such person was Paracelsus (1493–1541), a scholar, physician and alchemist who is widely credited with being the founder of modern pharmacy. Paracelsus championed the use of botany and alchemy to provide drugs and treatments for illness. He still believed that the Moon influenced disease and began to investigate the link. In his book *Diseases that Deprive Man of his Reason*, Paracelsus observed that madness or lunacy was greatest during the full Moon. His botanical remedies included 'lunar plants' such as Christmas rose (*Hellibora nigra*) and thyme (*Thymus majoram*). In terms of alchemy, where a substance's physical form and function were believed to dictate its actions, the Moon was linked to compounds of silver.[5] In these pre-Newtonian times the force of gravity was not understood, and the Moon was thought to exert its influence through magnetism. Paracelsus therefore also treated his patients with magnets and cold-water plunges.[6]

The public also believed that the Moon was an important influence on health. Elizabeth I wrote to her close friend the Earl of Leicester, 'Rob, I am afraid you will suppose by my wandering writings that a midsummer Moon hath taken large possession of my brains this month.'[7] The Elizabethans considered it wise to sleep in a darkened room in which moonlight was excluded, as it was believed that the Moon not only disrupted sleep but also caused rheumatic disease.[8]

An 18th-century apothecary's shop sign. The Moon was believed to influence human behaviour and represent cycles of life and death.

William Harvey (1578–1657), the physician who first described the circulation of the blood, likened circulation to that of the Sun and Moon, metaphorically linking the microcosm of man with the macrocosm of the heavens. The idea of circulation went against mainstream medicine and was at first resisted by the medical establishment. Only a few natural philosophers of the time such as René Descartes (1596–1650) saw any merit in Harvey's idea. To them the concept of circulation fitted well with the notion that we lived in a natural world where everything – celestial, political and economic – circulated around a central source – the Sun, the monarch and a capital city like London. This idea was best articulated by Robert Fludd (1574–1637), a mathematician and cosmologist who was a patient and friend of Harvey and eulogized his revolutionary ideas in *Medicina Catholica* (1629), writing:

Thus as the Moon follows her unchanging path, completing her journey in a month, she incites the spirit of the blood to follow in a cyclical movement. Every seaman is acquainted with the influence of the Moon on wind and tide. Why should she not exercise a similar influence in the 'microcosm' of Man?[9]

Through the eighteenth century there were no attempts to distinguish between different states of mental health. Thus if you suffered from epilepsy, sleepwalking, madness or demonic possession you were classified as a lunatic and if you were unable to look after yourself you were assigned to the lunatic asylum. At this time it was thought that those suffering from diseases of the Moon were more prone to episodic attacks at night, giving us the term 'moonstruck' from the German *Mondsüchtig*. This belief continued into the nineteenth century. It was thought,

Illustrations from Philippus Jacobus Sachse de Lowenheimb's *Oceanus macro-microcosmicus seu dissertation epistolica* (1664), linking the circulation of the cosmos, water and oceans with blood and the heart.

for example, that the waxing Moon heated the atmosphere, thereby melting the brain and provoking attacks of madness (or epilepsy). Others argued that epilepsy occurred during the new and full Moon when the planets and Moon were in specific locations and alignments.

As the nineteenth century advanced, progress in science and medicine increased rapidly, allowing the link between the Moon and madness to be explored using scientific methods. Although the advance in the long-term care of the mentally ill in asylums was a great step forward, the inmates were still treated with little respect. At the Bethlehem Hospital (known as Bedlam) in London, 'Bedlamites' were chained and flogged at certain phases of the Moon to prevent violence.[10] The number of inmates rose rapidly from a few thousand to over 100,000 in 80 years. Having large numbers of patients in one place and under the care of a few physicians or psychiatrists (as the new specialism was now called) allowed the first scientific studies to be conducted. The majority of these studies showed that there was no link between madness and lunar quartering. A notable 1854 study by Moreau, who observed institutionalized epileptics over five years, found no link with seizure frequency and lunar phase.[11]

Unfortunately this new information did not spread beyond psychiatric circles and was largely ignored by the public, and the general belief of 'lunar madness' remained. Indeed, the nineteenth-century judge Sir William Blackstone strengthened and formalized the link in English law in the Lunacy Act of 1890: 'A lunatic, or non compos mentis, is properly one who hath lucid intervals, sometimes enjoying his senses and sometimes not and that frequently depending upon the changes of the Moon.'[12] In English law this was not amended until the 1960s, and in the USA Congress did not vote to eliminate the word 'lunatic' from federal law until December 2012. Before this, u.s. legislation surrounding guardianship gave a bank the authority to act as a committee of estates on behalf of lunatics. This derogatory term was finally voted out by 398 to one.

In the 1960s and '70s, while the treatments for mental health in terms of care and medicine had advanced immensely, there

The case of the American seaman James Norris, who was shackled to his bed in London's Bedlam Hospital for a decade, stimulated the campaign for national lunacy reform.

was still no definitive consensus of any lunar effects. A survey of 29 published scientific studies during this period, covering psychiatric admissions, suicide and murder rates, shows that just over half (55 per cent) reported no influence of the Moon, while the remainder (45 per cent) reported some influence: six at full Moon, four at new Moon, one at first quarter (depending

99

on the Moon's orbit, whether at apogee or perigee) and one at all three phases – new, full and first quarter.[13] This disparate set of results did little to answer the question and only added to the confusion. If there was lunar influence on mental health, then how was it being manifested, especially as the effects were seen at different lunar phases? The water was muddied further by some studies speculating that the effect depended upon latitude (with the Moon's influence increasing closer to the equator),

The astronomer Maximilian Hell healing ailments by redirecting the forces of the Moon, 1780s.

syzygy (the alignment of the Sun, Earth and Moon) and/or planetary alignments.

The 'lunar effect' advocates at the time were very vocal, championing a tidal theory of lunar influence. They argued that just as the Moon affects the tides, it could influence human emotions, as we comprise 70 per cent water. The opposing lunar sceptics argued that the gravitational influence on such a small mass as a human is too minuscule to have any effect, pointing out that most lakes are immune to tidal changes. The sceptics also argued that the positive studies had not undergone the correct statistical analysis. In the 1980s Rotton and Kelly took all the available studies (both for and against) and amalgamated and examined them in a single study. This allowed a consensus to be drawn, ironing out the differences between the various methodologies used and making a statistical analysis clearer. The paper was finally published in 1985 and clearly found no links between lunar phase and mental health status: 'It is concluded that Lunar phase is not related to human behaviour and that the few positive findings are examples of type one error.' A 'type one error' is a statistical term which in this case means that the authors who reported lunar links were mistaken. In a less scientific tone, they stated: 'Just as we cannot prove that werewolves, unicorns and other interesting creatures do not exist, we cannot prove that the Moon does not influence behaviour.'[14]

After this the interest in studying the Moon for healthcare purposes evaporated, as it became extremely difficult to publish studies. Positive findings were considered statistically flawed and negative ones just confirmed what everyone already knew. This nadir lasted a decade or so before more statistically robust studies with larger numbers began to appear. In 2008 Martin Voracek and colleagues in Austria reviewed the timing of 65,206 suicides over 36 years in relation to lunar phases. This detailed study, one of the largest of its kind, found no relationship between suicide rates and lunar phases.[15] In another study Amaddeo and colleagues in South Verona, Italy, looked at the effect of the lunar phase on the demand for psychiatric services and after analysing ten years of case records found no links in over 8,000 patients.[16] A similar study of 771 participants from two Canadian emergency departments showed that psychological symptoms such as feelings of panic and anxiety were affected by season but not by lunar phase.[17]

Some have speculated that it is the brightness of a full Moon that triggers epileptic seizures indirectly by disturbing sleep patterns and sleep duration. Recent studies in healthy volunteers found that they reported being more tired in the morning following a full Moon, while another group took longer to fall asleep.[18] A study of epileptics found that seizure rates correlated with a full Moon but with cloud cover the effect disappeared, so it seemed that it was the brightness of the night sky that was responsible. Thus a bright night disturbs sleep patterns. This is demonstrated by the generalization that people who suffer from seizures and headaches tend to find that sleeping in darkened rooms provides them with better sleep.[19]

Most people living in urban areas are rarely exposed solely to natural moonlight. Electric light now predominates – a 100-watt bulb is 70 times brighter than a full Moon – so any lunar effects on mental health are masked in a modern society. The link between moonlight and mental health may have been more prevalent before the Industrial Revolution, especially when sanatoriums were situated in the countryside (away from public view) and inmates were exposed to moonlight and the brightened night sky, disturbing sleep patterns and possibly exacerbating mental illness.[20] Today the idea that moonlight influences our mood is enshrined in Yoko Ono's lyrics: 'Sleepless night, the Moon is bright'.

Lunar influence and surgery

Even today there is a belief (particularly in Germany and Switzerland) that there is a link between successful surgery and the Moon. The belief holds that during the waning Moon bleeding is more difficult to control, reducing the success of any surgery.[21] A study from Bavaria in 2003 of 866 operations in 782 patients found no relationship between poor outcome and lunar phase. In a 2001 study in Germany on patients recovering from surgery after breast cancer it was shown that the waxing or waning phases of the Moon did not affect their survival rate.[22] In another Bavarian study from 2009, in 2,411 lung cancer patients who had undergone lung surgery, post-operative morbidity, mortality and long-term survival were unaffected by the quartering of the Moon.[23]

Conflicting evidence about the lunar effect comes from two recent studies in the USA. The first, from Rhode Island, of 210

The association between death, grief and moonlight is depicted in this wood engraving by Thomas Bewick (1752–1828).

cardiac patients, showed that their recovery from surgery was shorter and chances of survival greater during the waning full Moon. In another American study of 3,935 patients, treated at Marshfield Institue in Wisconsin in 2012, it was found that the waning full Moon had no influence on the occurrence of acute cardiac problems.[24] Again the odds are stacked against a lunar effect and it is more the belief among patients and staff that influences the scheduling and outcomes of surgical interventions.

Change in renal function has been linked to the Moon phase, with a higher frequency of renal colic reported at full Moon.[25] In this case it is proposed that moonlight influences the body's melatonin levels. Production of melatonin is increased by light exposure, altering the endocrinological response of the body.[26]

The Moon and childbirth

For many years it has been a widely held maxim among midwives that delivery rates, births (natural versus induced) and complications are influenced by lunar phase. This is an extension of ancient folklore that lunar phase influences fertility. However, a 2005 analysis of over 560,000 births in North Carolina showed that the Moon had no influence on the delivery of babies.[27] These findings were confirmed in a 2008 German study of 6,725 births which also found that gender was unaffected by the Moon.[28] The authors of the 2005 American study speculated that perhaps midwives' belief in lunar influence was so ingrained that they concentrated only on negative experiences during a full Moon, thereby reinforcing the belief. Even in assisted conception and in vitro fertilization the lunar phase exhibited no influence in a 2005 study from Liverpool, which analysed results collected over thirteen years from 721 women.[29]

Intriguingly, a 2012 study showed a monthly relationship in a semi-rural population in Israel.[30] This study of over 800,000 live and stillbirths in Jewish and Muslim groups found a significant but complex relationship to the lunar month and time of year, with live birth rates peaking during a full Moon. The subjects live in an area where artificial light is less invasive, so any lunar effects may be a reflection of this rather than gravitational forces. A 2012 Brazilian study of 17,417 births showed that atmospheric conditions or weather affected the birth rate more than the lunar phases.[31]

Up until the twentieth century birth defects were also thought to be caused by the Moon, or at least by too much exposure to moonlight. This idea is reflected in the now obsolete term 'mooncalf', a medical term used to describe a hard, swollen piece of tissue in the womb, like a fibroid or uterine mole, often mistaken for pregnancy.[32] The term was also used to denote an imbecilic person. In Shakespeare's *Tempest*, Caliban is referred to as a mooncalf and described as 'a freckled whelp, hag-born – not honour'd with / A human shape' (i, ii). In Charles Dickens's *Great Expectations* (1861), when Pip, the young protagonist, is given the opportunity to pay a visit to a potential benefactor, Miss Havisham, his older sister, Mrs Gargery, gets excited and flustered, stating, 'Here I stand talking to mere Mooncalfs . . . and the boy grimed with crock and dirt from the hair of his head to the sole of his feet.' A common belief held until the twentieth century was that a deformed baby might be blamed upon Moon-blasting – a belief that the pregnant mother had been exposed to excess moonlight, for instance by travelling in the moonlight.

Throughout history there has been no shortage of vociferous writers depicting the positive lunar influences on our health. While it seems that any lunar effects are minor and largely a result of lighter skies, it is possible to exclude the influence of human superstition and belief. Support by connection sounds convincing; thus, if solar storms alter the magnetosphere, and this alters human mental health, then as the Moon influences the strength of solar storms, the Moon must also influence human mental health. The ability to attribute altered mental health to subtle lunar effects is made more difficult by the fact that the boundaries between distinct psychological disorders have become more and more blurred as we learn more about their aetiology. Nowadays this distinction is blurred further by the widespread use of medication in treating illness, blunting even more any potential lunar biological effects.

Throughout history we have speculated on lunar effects on health but after two centuries of close scientific scrutiny, reported in dozens of studies, no credible link between the Moon, its phase orbital position, and alignment with the Earth and Sun and our health can be demonstrated. It is more likely to be human belief that fuels this association. With modern living, pharmacological treatment of mental health conditions, the refinement in surgical technology and midwifery, any subtle lunar effects are likely to be swamped. Modern times have broken our primeval link with the Moon, our most ancient of partners.

5 Lunar Art and Literature

The Moon is widely represented in literature and art, and has inspired writers and artists from ancient times; there are hundreds of poems, plays and stories describing the Moon's various appearances, both physical and metaphorical. Virtually every well-known writer has something to say about the Moon or has used its characteristics to set the scene, whether it is a romantic moonlit situation, a ghostly dark Moonless night or a supernatural full Moon.

The Moon as a planetary body or simply an extraterrestrial utopia has played a major role in the development of the science fiction genre, particularly in Victorian literature. As fiction writers and scholars first imagined and speculated about the Moon, its physical characteristics and whether human-like civilizations existed on or below its surface, this influenced the development of astronomy, selenology (the scientific study of the Moon) and even aeronautical engineering. The Moon has featured in hundreds of science fiction stories and films. These can be broadly split into those where the Moon is seen as a harsh colonial outpost where strange things happen, such as alien encounters, and those where the Moon provides an environment for the founding of new utopian societies. These writers explore contemporary fears and hopes of a technological future, along with the associated angst of Earth's present philosophical and political systems.

One of the earliest (first century AD) works of science fiction is *De facie quae in orbe lunae apparet* (Concerning the Face which Appears on the Orb of the Moon) by Plutarch. While the first part of the work summarizes contemporary lunar knowledge, the second part delves into pure speculation, describing a Moon inhabited by intelligent humanoid creatures living on a verdant – owing to the Moon's moisture-promoting properties – world. The Roman Cicero (106-43 BC) informs us that 'Xenophanes says that the Moon is inhabited, with a country having several towns.'[1]

The earliest-known fictional work describing a voyage to the Moon was by the Greek Lucian of Samosata in AD 160. Originally not intended as a work of science fiction but as a farce, it is called *True Story*. Even today it provides an enjoyable read. It tells the tale of Greek explorers who set sail on the Atlantic, where, during a severe storm, their ship is whisked into the air by a waterspout (not unlike the tornado in *The Wizard of Oz*). After an eight-day voyage they end up on the Moon. The Moon is populated by the Moonites, who are at war with the Sunites, who live on the Sun. The two groups are fighting over ownership of the morning star. Eventually the Sunites win by creating a permanent eclipse of the Moon, forcing the Moonites to sign a peace treaty in which it is agreed that the morning star remains neutral ground and unoccupied. The king of the Moon, Endymion, invites our adventurers to stay and offers the Greek leader his son in marriage as there are no women on the Moon. It seems that the Moonite men act as wives until they are 25, at which time they become husbands. Lacking wombs, gestation rather bizarrely occurs in the calf muscle. The adventurers politely decline this offer and leave to continue with their adventures.

Johannes Kepler (1571–1630), the mathematician and astronomer, was interested in the Moon throughout his life and even had a copy of *True Story*. Kepler spent most of his academic life studying the movement of planets, which led him to propose that it was elliptical orbits that best explained the observations of the planets, marrying together Aristotelian and Copernican cosmology. This major scientific finding secured Kepler's place in history as a great scientist. He was the first scientist to explain that the tides were caused by the Moon. In Prague in 1609, while employed as the imperial mathematician, Kepler was asked by Emperor Rudolf if he believed that the patterns on the lunar surface were merely reflections of the Earth's landforms, as many believed at the time. In private he speculated about what an observer on the Moon would experience, which led him to think about travelling to the Moon. His ideas were not published in his lifetime as they were unfinished and, more importantly, he was afraid that he and his mother would be accused of sorcery or witchcraft. The book *Somnium sive astronomia lunaris* (The Dream) was published posthumously in 1634 by his son. In the story the principal character, Duracotus, and his companions travel to the Moon, known as Lavania and inhabited by the Privolvans, during a solar eclipse: a time when a 'bridge of darkness' links the Earth and the Moon. The book explores Kepler's understanding of gravity in space and his prediction that on an extraterrestrial location like the Moon one would be weightless. This remarkable book was originally published in Latin and was lost to obscurity for over 260 years until in 1898 a German translation was released and a wider readership was able to appreciate its fictional value.

Selenographia; or, Newes from the World in the Moon to the Lunatics of this World was written by Lucas Lunanimus of Lunenberg (1699) and describes his alleged trip to the Moon using a kite as transport. In 1638 *The Man in the Moone; or, a discourse of a voyage thither* was published posthumously by the bishop of Hereford, Dr Francis Goodwin (1562–1633). The lunar traveller this time is carried to the Moon by captive swans, taking twelve days to arrive. He finds the Moon covered by a vast sea interspersed with small islands. It is portrayed as a utopia where the ordinary people, the Lunars, are ruled by a benevolent monarchy, with no crime or illness and ground so fertile that food is plentiful, allowing the labouring classes to lead a more carefree life.

In 1783 the Montgolfier brothers invented and popularized the hot air balloon, providing for the first time a plausible way to fly. With little known about the limits of atmosphere the sky was open to exploration both physically and in the imagination – in fact the 'sky was the limit'. This inspired writers of science fiction and fantasy. One work, *The Travels and Surprising Adventures of Baron Munchausen* by the scientist Rudolph Eric Rapse (1736–1794), was published in 1793. In this tale a balloon expedition proceeds to the Moon, where the baron finds his old associate Berthold. Eventually the Baron angers the king of the Moon, who resents him for his romantic connections with the queen of the Moon. A bungled escape brings the Baron back to (and beneath) the Earth.

In 1835, in *The Unparalleled Adventures of one Hans Pfaall* by Edgar Allan Poe (1809–1849), Hans Pfaall flees his creditors

A fictional flight to the Moon from Roman de Gonzales' *L'Homme dans la lune* (1648).

By means of clambering up a turkey bean, Baron Munchausen recovers his silver casket, which had been bounced up to the Moon.
A chromolithograph from a French edition of his *Travels and Surprising Adventures*, *c.* 1850.

by ballooning to the Moon. The story has an interesting opening with the arrival in Rotterdam of a frightened balloon-borne alien who, before returning skywards, delivers a package addressed to the 'Excellences of the States College of Astronomers'. This contains the manuscript from Hans describing his journey to the Moon and his subsequent five years living with its mute inhabitants, who cannot use sound to communicate because of the rarity of the atmosphere. The story tries to be scientific and addresses the issue of travelling at extreme altitudes, but does not mention how the balloon transfers from a terrestrial to a lunar orbit. This is one of the first works of fiction that offers us the view that Selenites may be alien life forms and not human-like.

The majority of early Victorians believed that life, intelligent or otherwise, existed on the Moon. Telescope technology was making great progress, with ever-larger telescopes being constructed across the world. It was hoped that it would only be a matter of time before they were able to achieve magnifying power sufficient to see the lunar surface close up. While Victorian observers could see small craters, mountains and valleys, at a fairly high resolution the lunar surface still looked dry, grey and lifeless, without any obvious vast oceans. It was known by observing star occultation that the Moon did not have an atmosphere or weather system. The Moon's mass had also been calculated, and found to be lighter than expected. All these observations drove speculation that the Moon was hollow and contained an ocean inside. Most scientists concluded rightly that life as we knew it was impossible on the Moon's surface but they could not disprove the possibility that life existed in a subterranean form, especially if there was a subterranean ocean. Below the surface the extremes of diurnal temperature could be tolerated. It was also conceded that there might be a very narrow atmosphere clinging to the surface which telescopes would not have been able to pick up. Although it would not be human-like, many speculated that intelligent life was probable. Strangely, no one imagined that these lunar inhabitants would be more sophisticated than us or want to communicate with the Earth. While we now know that no subterranean oceans exist, lunar caves and hollow lave tubes do and are thought to offer some of the best places for establishing a Moon base, as being below the surface they would shield occupants from the extremes of temperature as well as from solar and cosmic radiation.

One of the most famous hoaxes in the history of journalism concerns supposed life on the Moon. Although intended as a work of non-fiction, and now known as the Great Moon Hoax of 1835, it is with hindsight a work of pure science fiction. Originally it appeared in a series of six daily articles, published between 25 and 31 August under the title *Great Astronomical Discoveries lately made by Sir John Herschel at the Cape of Good Hope*. The *New York Sun* reporter Richard Adam Locke wrote the articles simply to increase the paper's circulation, while at the same time mocking other more serious observers who had hinted that they had evidence to suggest the Moon was inhabited.

The first article sets the scene, describing the new advances made in telescope technology by Sir John Herschel (who was a real astronomer of some fame). The article reports the building of a new 7-ton telescope with a magnifying power of 42,000 times, capable of resolving objects on the Moon's surface just 40 cm across. The news is reported by his fictitious assistant, Dr Grant.

The second article describes this new instrument in great detail, giving the reader enough scientific terminology to make it believable. Halfway through the article the telescope is turned on the Moon and slowly the facts are drip-fed to the unsuspecting reader. The very first things to be seen are hexagonal basaltic rocks covered in red poppies: 'this was the first organic production of nature, in a foreign world, ever revealed to the eyes of man'. Having proved that the Moon not only has an atmosphere but also life, the article quickly moves on, describing great mountains, volcanoes, white beaches and vast lakes and oceans 'stretching for hundreds of miles'. Next a forest of oak-like trees appears and in its shaded interior are herds of 'brown quadrupeds' resembling bison. Other creatures are described – a giant goat-like creature, some grey pelicans and a black and white crane. Cloud eventually blocks the view and the article closes with hints of more fantastic discoveries to come. This report in such a reputable paper set the world's media alight. The third instalment provides even more revelations about lunar life, reporting that Dr Herschel had identified 38 species of tree and nine new species of mammal, including reindeer, elk, bear, pygmy zebra and, most fantastic of all, a beaver that walked on two legs. These revelations continue the next day: after seeing lunar-giraffes, and improving the resolution of the telescope further, Dr Herschel and his assistant observe humanoids – not like those on Earth but winged like bats. These humanoids are around 1.2 m tall and yellow-skinned with features more human than simian. The paper names the new species *Vespertilio homo* and claims that their discovery was originally going to be kept secret. The next instalment reports the discovery of uninhabited buildings reminiscent of Greek temples. In the sixth and final instalment the bat people are back in the observer's field and this time they are seen socializing, eating, bathing and generally enjoying themselves. No one appears to be at work or industry. With observation time over, the astronomers go off to bed, but are soon woken by a commotion, with the observatory ablaze. While the fire is soon extinguished and the telescope repaired it is never again trained on the Moon. With each new revelation the paper's circulation soared. It was soon republished as a pamphlet and syndicated to other newspapers around the world. The story struck a chord with the public, especially the idea of the Moon as a paradise – not lost but to be obtained. (During this period North America was undergoing colonization on a massive scale.).[2] Amazingly enough it was several weeks before the hoax was revealed, or at least admitted to. Strangely Sir John Herschel did not deny the validity of the claims, as he thought it a trivial matter. A modern twist to the hoax is that Neil Armstrong, the first person to walk on the Moon, died on the 177th anniversary of its publication.

Following the success of the hoax a rival publication, the *New York Transcript*, republished Edgar Allan Poe's earlier Hans Pfaall story under the headline 'Lunar Discoveries, Extraordinary Aerial Voyage by Baron Hans Pfaall'. Although it contained some scientific facts the report was never really accepted as anything other than fantasy.

Claims of life on the Moon persisted until the Russian probes reached it in the 1960s. While scientists were in agreement that the Moon could not support large animals, it was claimed by a Professor Pickering, a well-respected astronomer of his time, that he had seen evidence of large insect swarms on its surface. While observing between 1919 and 1924 from his personal observatory in Mandeville, Jamaica, he reported seeing patches on the lunar surface change in size. At first he attributed these to snow and ice, and later vegetation, but he then reported that the 'patches' changed location, and that 'the distance involved is about twenty miles and completed in twelve days . . . which as we have seen implies small animals'.[3] These observations of 'lunar insects' were widely published, owing to Professor Pickering's standing as a respected astronomer, but no one took the reports seriously or refuted his detailed observations. Pickering is another person who has a lunar crater named for him. By the 1960s the idea that life as complex as insects could exist on the Moon was no longer accepted, but life on the Moon was still possible, and it was speculated that the patches represented simple vegetation or microorganisms that expanded and contracted with the extremes of lunar temperatures.[4]

The modern writing genre of science fiction begins with Jules Verne and H. G. Wells. Jules Verne (1828–1905) published *De la Terre à la lune* (*From the Earth to the Moon*) in 1865. This is the first modern fictional work about a trip to the Moon based on scientific principles and even today provides a good read. Jules Verne used calculations made by his brother-in-law who was a professor of astronomy. He introduced to the general public several issues about space travel such as the fact that a vehicle leaving the Earth would have to exceed 40,000 km per hour (known as the Earth's escape velocity), collisions with meteorites and steering using directional rocket propulsion. The tale's protagonists, the Baltimore Gun Club, decide to build a 270 m-long cannon called *Columbiad* and, using 723 tonnes of explosive, blast a hollow shell to the Moon. The shell or spaceship accommodates three travellers, a variety of equipment and a dog on a one-way trip. *Columbiad* fires the projectile into space but misses the Moon, allowing the astronauts to circumnavigate the Moon and return to Earth safely, where they land in the sea.

Many aspects of Verne's story came true with the Apollo programme – a Florida launch site and splash down in the sea, for example. The Apollo 11 command module was named *Columbia* after the fictional spaceship. The retired judge Jean Jules Verne (Verne's grandson) personally witnessed the launch of Apollo 8 in 1968, recounting to reporters that he remembered his grandfather telling him he would see men going to the Moon.[5]

Another influential book of this era, *The First Men in the Moon* (1901), was by H. G. Wells. It tells the tale of two travellers (one a scientist, Mr Cavour, and the other a young businessman

Jules Verne in his novel *From the Earth to the Moon* believed the Moon had its own breathable atmosphere.

Mr Bedford) who journey to the Moon. Travel is made possible by the development of a revolutionary new metal that has anti-gravity properties. They pilot their spacecraft by directional control of the new metal, landing on the Moon just before dawn, when the surface temperature is so cold that the lunar atmosphere is frozen. However, during the day, as the sunlight returns, the surface bursts forth with an abundance of vegetation. This is consumed by giant herbivorous beasts shepherded by intelligent man-sized ant-like creatures that live below the lunar surface in vast caverns encompassed by a vast internal ocean. The book ends with the inventor of the propulsion system remaining stranded on the Moon while the young man returns to Earth. The stranded scientist transmits a series of messages back to the Earth, describing his life among this new society. Eventually these messages cease and no further use of the metal is made.

As science fiction became popular in the twentieth century, hundreds of Moon-related stories were written. Virtually every sci-fi writer from Asimov to Wyndham has written one. The most prolific authors, Arthur C. Clarke and Robert Heinlein, explored the Moon as a utopia/dystopia, a venue for alien encounters and the exposition of futuristic technology and colonization.

From its beginning, the medium of film adopted the genre of science fiction. One of the earliest science fiction films was the French silent movie masterpiece, *Le Voyage dans la lune* (A Trip to the Moon) (1902), directed by Georges Méliès. Only fourteen minutes long, it is broadly based on Verne's *From the Earth to the Moon*. Although it is more fantasy than science, it is important in setting the scene for future lunar science fiction films.

A still from Georges Méliès's *A Trip to the Moon* (1902).

A film that tried to be technically accurate was *Die Frau im Mond* (The Woman in the Moon, 1929), by the German director Fritz Lang (1890–1976). Lang, who had made *Metropolis* (1926), was keen to make his next sci-fi film scientifically accurate so he hired Hermann Orbeth (1894–1989), one of the early pioneers in rocket design, as consultant.[6] While a rocket is used to reach the Moon, the film has little realism. The lunar surface looks more like the Sahara desert than anything. While Orbeth informed Lang that the Moon was airless and the characters would have to wear spacesuits, Lang replied, 'How could one present a love story taking place on the Moon and have the lead characters talk to each other and hold hands through space suits?'[7] Although disliked by the critics, it was Germany's most popular film that year.[8] During the Second World War, the Gestapo destroyed every copy of the film they could find as they thought the rocket imagery was too realistic and would reveal how their own v2 rockets operated.[9]

After the success of the v2 rockets in the Second War World, rocket travel to the Moon seemed both feasible and probable. This optimism is expounded in the Oscar-winning film, *Destination Moon* (1950), directed by Irving Pichel (1891–1954). Again, based broadly on the Jules Verne tale, a group of industrialists come together and decide that the only way to protect the free world is to bring the Moon under Western influence. The rocket, though nuclear-powered, has a single stage with all four astronauts travelling to the Moon in a small cabin at the top. The Moon is shown as desolate and rocky, populated with steep ridges and high-pointed mountains formed into long ranges. The mountain peaks were perceived this way as their projected shadows viewed by terrestrial telescopes looked jagged. The film ends with the astronauts safely returning to Earth.

To the 1950s cinema-goer this became the popularized view of the Moon and was repeated in *Radar Men from the Moon* (1951), *Project Moon Base* (1953), *Catwomen of the Moon* (1953) and its remake *Missile to the Moon* (1959). This view was also paralleled in children's stories, in two of the famous graphic novels created by the Belgian cartoonist Georges Remi (1907–1983), featuring

A poster advertising the science fiction film *Destination Moon* (1950).

TWO YEARS IN THE MAKING!

DESTINATION MOON

COLOR BY TECHNICOLOR

Produced by GEORGE PAL · Directed by IRVING PICHEL
Screenplay by RIP VAN RONKEL, ROBERT HEINLEIN and JAMES O'HANLON

Hergé's 'Adventures of Tintin', *Destination Moon* and *Explorers on the Moon* (both from 1959). In these comic novels the Moon rocket is a single stage rocket, an exact copy of the rocket in the film *Destination Moon*. In this two-volume adventure Snowy is the first dog on the Moon and even has his own spacesuit. While exploring the Moon, Tintin and Snowy discover a cave complex with a floor of ice. The lunar imagery, gadgetry and storyline of these cartoons provide a wonderful snapshot of the post-war world. A great homage to this era can be seen in the Wallace and Gromit cartoon *A Grand Day Out* (1989) directed by Nick Park (b. 1958). This award-winning, twenty-minute piece is a tour de force of our pre-Apollo view of the Moon. Wallace, a cheese

The dried mud in the foreground gives the lunar surface an arid feel, from the film *Destination Moon* (1950).

connoisseur, and his dog Gromit travel to the Moon in their homemade rocket to sample the delights of lunar cheese.

The 1960s continued with a romanticized and exaggerated view of the Moon. In *The First Men in the Moon* (1964), directed by Nathan Juran (1907–2002), a United Nations spacecraft lands on the Moon only to find a Union Jack that had been planted in 1899, with a letter proclaiming the Moon for Queen Victoria. The Moon is inhabited by civilized humanoid insect-like Selenites, and follows Wells's story. In 1967, when the Apollo programme was in full swing, the film *Countdown*, directed by Robert Altman, was able to show the complex preparations that were needed for a lunar mission. Despite this, once the astronaut has landed on the Moon, the lunar surface is shown with craggy mountains, a view of the lunar landscape from the 1950s. *Mission Stardust* (dir. Primo Zeglio, 1968) is similarly unrealistic. Realism finally appeared in 1969 with the film *2001: A Space Odyssey*, directed by Stanley Kubrick (1928–1999) and adapted from the book of the same name by Arthur C. Clarke. It is now credited with being one of the greatest science fiction films of all time. The story depicts one of the most famous fictional alien artefacts: two black mono-liths are found, one on Earth and one on the Moon buried in the crater Tycho, transmitting a radio signal towards Jupiter. The monolith's discovery is kept secret by the Western nations so that they can remain ahead of the Eastern powers such as the USSR. The depiction of the lunar surface was, for the first time, realistic, and the film received an Oscar for its visual effects. The filmmakers knew what the surface looked like, since the Russian spacecraft Luna 9 had landed on the Moon in February 1966 and the American Surveyor in June, and both had sent back pictures of the surface. The view of craggy mountains was gone for good.

NASA has thousands of close-up images of the lunar surface taken by their unmanned space vehicle the Lunar Reconnaissance Orbiter, launched in 2009. These are available for public view-ing on their website. With *A Space Odyssey* in mind, they have asked the public to look through the images for alien artefacts. The lunar surface is a good place to look as any evidence should remain undisturbed for many millennia. Finding such evidence would alter human culture for ever. The images of the Apollo landing sites clearly show the tracks left by the astronauts remaining unchanged for 50 years.

After the Apollo landings, apart from the TV series *Space 1999* and the British children's programme *The Clangers*, media interest in the Moon waned as it had been shown to be a cold, lifeless world. This desolate view is reflected in the Duncan Jones film *Moon* (2009), where the sinister plot unfurls within the confines of a mining facility on the far side. *Apollo 18* (2011) directed by Gonzalo López-Gallego is a sci-fi horror film and uses the cancelled Apollo 18 mission to provide the backdrop for an alien contact story. Other recent films, such as the epic *Iron Sky* (2012) and *Transformers 3: The Dark of the Moon* (2012) explore similar concepts with the Moon. In the disaster film *Impact* (2009), directed by Mike Rohl, the Moon is hit by a brown dwarf fragment. The resulting heavier Moon then begins to spiral towards the Earth. In typical disaster movie style, the scientists in a just few weeks resolve the problem and miraculously save the Earth with only a few hours to spare.

The Moon in art

The Moon is a popular subject for artists worldwide, and in most national art galleries there can be found at least one work showing a Moonlit scene. In Western art most works depict actual places: exotic locations or mundane towns, such as J.M.W. Turner's *Moonlight at Millbank* (1797). Scenes are sometimes dark, sometimes bright and covered with a silvery wash of moonlight. Another popular subject is the Harvest Moon, and paintings featuring this often show rural and agricultural scenes. Notable works by more recent artists include *Evening Landscape with Rising Moon* (1889) and *Country Road in Provence at Night* by Vincent van Gogh (1853–1890), the latter one of many showing the Moon as simple curved splashes of yellow. In *Moonlight* (1926) by Salvador Dalí, the Moon is hardly visible at all. Edouard Manet (1832–1883), in *Moonlight over Boulogne Harbour*, paints a dark and chimerical scene, with the Moon shown shining

Edouard Manet, *Moonlight over Boulogne Harbour*, 1868.

J.M.W. Turner, *The Fighting Temeraire*, 1838, with a waxing crescent moon above the old man-of-war.

Henri Rousseau, *Carnival Evening*, 1886.

through the clouds and casting shadows on the quayside and on the water. Henri Rousseau (1844–1910) included the Moon in many of his paintings: in *The Snake Charmer* (1907), the Moon adds a dreamlike atmosphere to the jungle scene, while in the more cheerful *Carnival Evening* (1886) and *The Sleeping Gypsy* (1897) a full Moon is shown with a smiling face.

In Eastern art the Moon is used to illuminate night scenes but is more often included to symbolize emotional, religious or fictional topics. In Japan and China the inclusion of the Moon provides a spiritual meaning while setting the season of the work. A good example is the nineteenth-century work of the Japanese artist Tsukioka Yoshitoshi (1839–1892) who produced a series of 100 woodblock prints called *One Hundred Aspects of the Moon*. In Indian art the Moon is used simply to show that a night scene is being depicted. The most popular imagery is that showing

romantic trysts. Notable examples from the twentieth century are from the Japanese artist Ohara Koson (1877–1945) who produced a distinctive series of images depicting close up views of water fowl with the full Moon in the background.

The Moon is also used widely in commercial art and appears in many company logos. One of the most famous is for Dreamworks animation, which shows a boy fishing while sitting on the crescent Moon. The UK hotel chain Premier Inn has on the front of all of its hotels a logo of a smiling man in the crescent Moon. The Moon here represents a good night's sleep. For over 40 years the logo for Proctor & Gamble showed a bearded man in a crescent Moon, until the logo was changed to the company's initials in the 1990s. The change came about because many thought the previous logo was full of satanic symbolism and even included the secret inclusion of '666' in the man's curly beard.

A Japanese porcelain plate (*c.* 1900) with a full Moon and bamboo motif.

A 1920s advert for Lever Brothers Ltd's domestic cleaner Vim.

A public health advert of 1989, with an owl and the full Moon providing a sinister backdrop.

AIDS IS NOT A QUICK KILL

The owl represents prophecy and is also an omen of prolonged illness or death.

American Indian Health Care Association

A well-known
hotel chain logo.

Moonlight and the novel

In his novella *Strange Case of Dr Jekyll and Mr Hyde* (1886), Robert
Louis Stevenson (1850–1894) explores the duality of personal-
ity, in which the virtuous Dr Jekyll, physician and scientist, is
transformed into the evil Mr Hyde. After taking his potion Mr
Hyde's most depraved act is the murder of the MP Sir Danvers
Carew. The murder, lit by a full Moon, is witnessed by a maid
from her bedroom window. Stevenson does not comment on
whether it is the full Moon that drives the evil Mr Hyde to this
ultimate act of depravity or whether it is a means by which the
maid identifies the murderer. It is most likely both.

Moonlight helped inspire Mary Shelley (1787–1851) to write
Frankenstein (1818): the idea for the story arose when she awoke
from a deep sleep and found moonlight flooding her room. Harper
Lee (b. 1926), in her 1960 novel *To Kill a Mockingbird*, uses the
Moon to infuse the narrative with a 'spooky' atmosphere whenever
the young protagonists, Scout, Jem and Dill, decide to investigate
their elusive neighbour Boo Radley. In *Death on the Nile* (1937)
by Agatha Christie (1890–1976) the obligatory murder takes place
on the deck of a Nile cruiser on a moonlit night. From the same

year, *The Hobbit* by J.R.R. Tolkien features the Moon prominently throughout. Here the Moon is not only important in terms of
setting the degree of nocturnal illumination and atmosphere
and mood, but also in contributing to the mythical folklore of
Middle Earth. 'Moon letters' (writing only visible during the
same lunar conditions they were written under) provide the
central clue to gaining access to the back door of the lair of Smaug,
a fearful dragon. Later, when the annoyed Smaug leaves his lair
to destroy the lake town of Esgaroth, it is the Moon illuminating
the dragon's underbelly that allows the archer, Bard, to shoot his
last remaining arrow into Smaug's only unprotected spot, killing
him. Freeing the folk of Middle Earth from tyranny and allowing the hobbit Bilbo Baggins to return home, the scene is set for
the later 'Fellowship' trilogy.

George Eliot (1819–1880) used folklore in her novel *Adam
Bede* (1859), about the quartered Moon influencing rainfall: when
the arms of the waning crescent Moon are turned up, the Moon
acts as a basin and collects water, so the weather will be dry; but
during the waxing crescent Moon, the crescent points to the
Earth and the water therein is lost and rain follows. Thus during
the burial of Thias Bede: 'They'll ha'putten Thias Bede i' the
ground afore ye get to the churchyard. It 'ud ha' been better luck
if they'd ha' buried him i' the forenoon when the rain was fallin';
there's no likelihoods of a drop now, an' the Moon lies like a boat
there, dost see? That's a sure sign o'fair weather; there's a many
as is false, but that's sure.'

Charles Dickens referenced lunar folklore in a similar way in
David Copperfield (1849), where the common association between
the Moon, the tides and water is used to great effect during the
death of Mr Barkis:

> 'He's a going out with the tide,' said Mr. Peggotty . . .
> I repeated in a whisper, 'With the tide?'
> 'People can't die, along the coast,' said Mr. Peggotty,
> 'except when the tide's pretty nigh out. They can't be born,
> unless it's pretty nigh in – not properly born, till flood. 'He's
> a going out with the tide. It's ebb at half-arter three, slack

The murder scene at full Moon in Robert Louis Stevenson's *Strange Case of Dr Jekyll and Mr Hyde* (1886).

TRAMPLING HIS VICTIM UNDER FOOT, AND HAILING DOWN
A STORM OF BLOWS

water half an hour. If he lives till it turns, he'll hold his own till past the flood, and go out with the next tide.'...

Mr Peggotty touched me and whispered with much awe and reverence 'They are both a-going out fast.'...

'Barkis is willin'!'

And, it being low water, he went out with the tide.

European writers are just as likely to wax lyrical about the Moon. In *The Divine Comedy: Paradise*, Dante Alighieri (1265–1321) on his trip through the heavens visits the Moon, while Cyrano de Bergerac in his *Histoire Comique de la lune* (A Comic History of the Moon, 1657) travels to the Moon to meet the Lunarians. The German writer Johann Wolfgang von Goethe (1749–1832) often mentions the Moon in his works, but solely in terms of the illumination it provides, as in 'Melpomene': 'Down from the heavens the Moon at her full was shedding her splendour.'

Shakespeare and lunar folklore

For William Shakespeare (1564–1616) the Moon provided opportunity for great drama, and dialogue for his characters both historical and fanciful. Many lunar references abound in Shakespeare's works, and he alludes to virtually every superstition, folklore, cultural belief and physical characteristic of this argentine orb extant at the time. Apart from the obvious references to the phases of the Moon or the silvery colour of moonlight, Shakespeare mostly utilizes the Moon as a portent of bad tidings, particularly an eclipse. In *Sonnet 107*, 'The mortal Moon hath her eclipse endured, / And the sad augurs mock their own presage', refers to the death of Elizabeth I as Cynthia, astrologers and fortune tellers are proved right in their predictions of her imminent death. This, matched with the belief that witches were able to take demonic power from the Moon, provides the Bard with some powerful imagery. While brewing a magical potion, one of the three witches in *Macbeth* chants, 'Gall of goat, and slips of yew / Silver'd in the Moon's eclipse' (IV, i). The mooncalf Caliban from *The Tempest* is an example of lunar madness as a useful tool, along with dogs and wolves howling at the Moon. In *Othello* the Moon brings madness and death to both Desdemona and Othello.

The Moon's effect on the tide and its connection with water is alluded to in several plays. In *Richard III* (II, ii), Queen Elizabeth has the line, 'That I, being govern'd by the watery Moon / May send forth plenteous tears to drown the world', while in *Love's Labour's Lost* (v, ii), the king says, 'Vouchsafe, bright Moon, and these thy stars to shine, / Those clouds removed, upon our watery eyne.'

We can find reference to the chaste classical female Moon deity, Luna or Diana, a late Tudor and Jacobean phenomenon, as Romeo and Juliet swear by the Moon (ii, ii):

Romeo: Lady, by yonder blessed Moon I swear
That tips with silver all these fruit-tree tops –
Juliet: Oh! swear not by the Moon, th'inconstant Moon,
That monthly changes in her circled orb . . .

A Midsummer Night's Dream provides a cornucopia of lunar metaphors and references as the play is set under a moonlit sky, has a character called Moonshine in the play within a play and is full of fanciful fairy-like characters. The Moon influences the

A parody of the three witches in Shakespeare's *Macbeth* is shown in this political cartoon of three politicians (with William Pitt the younger in the middle), 1791.

To H: Fuzelli Esq! this attempt in the Caricatura-Sublime, is respectfully dedicated.

WIERD-SISTERS; MINISTERS of DARKNESS; MINIONS of the MOON."
'They should be Women,' – and yet their beards forbid us to interpret, – that they are so.'

Théodore Roussel,
Moonrise from the River
(1912). A softground
etching and aquatint
depicting the River
Thames at Abingdon
in Oxfordshire.

behaviour and reactions of the characters, who symbolize love, lust and dreaming. *A Midsummer Night's Dream* and *Twelfth Night* are thought to be influenced by a great event Shakespeare witnessed as a young boy in 1575. It was at Kenilworth Castle, where Robert Dudley, the earl of Leicester, was holding a grand fete for Queen Elizabeth. He was to use the event as his last great attempt to gain her hand. The earl arranged a number of spectacular water pageants (Kenilworth Castle was surrounded by water on three sides) filled with classical symbolism and breathtaking fireworks. The earl was hedging his bets: the widow Lady Douglas had already given him a son, but as the marriage had been conducted in secret the earl still felt free to look for a better prospect. During the celebrations the queen found out about the marriage to Lady Douglas and left, losing all interest in the earl as a husband. Soon afterwards he married the widow Lettice Knollys, one of the queen's closest friends. In *Twelfth Night* and *A Midsummer's Night Dream*, Shakespeare combines this pageantry with farce to great effect, where Cupid (Dudley) fails to win the heart of the chaste moon goddess (Elizabeth), his arrow instead piercing a little white flower, transforming it into 'love-in-idleness' – another name for the pansy, representing Lettice.[10]

Allen William Seaby, *Nightingale*, 1920s, woodcut.

Film and the Moon

As well as books and plays, there are many films with references to the Moon in their title. Since the 1930s there has been a film with 'Moon' in the title released virtually every year, with a lull during the 1940s during the Second World War. (Surprisingly there was also a lull in the 1970s, at the time of the Apollo landings. These were heavily promoted on u.s. television and soon the public became a little tired of anything to do with the Moon.) Many have geographical titles, linking a romance with an exotic or everyday location, such as *Moon over Miami* (1941), *Montana Moon* (1930), *Moon over Parador* (1988), *Alabama Moon* (2009), *Roll on Texas Moon* (a cowboy film from 1946), *Under the Hula Moon* (1995) and *Under the Pampas Moon* (1935). Films with quirky names like *Sky Full of Moon* (1952), *A Moon for the Misbegotten* (1960), *Nor the Moon by Night* (1958) and the *Three-cornered Moon* (1933), a comedy about the stock market crash starring Claudette Colbert, use the contemporary view of the Moon's ability to influence our thoughts and behaviour, and its association with the supernatural. Another notable title is *Paper Moon* (1973), directed by Peter Bogdanovich, which, like the song of the same name performed by Ella Fitzgerald and Nat King Cole, is a metaphor for imitation and false hope. Perhaps the best-known cinematic lunar quote is from one of Hollywood's most famous actresses, Bette Davis, and is the last line of the film *Now, Voyager*: 'Jerry, don't let's ask for the Moon – we have the stars.'

The musical Moon

While there are hundreds of films and novels with lunar-related titles, there are just as many songs containing Moon lyrics. Apart from 'Moon' being an easy word to rhyme, it is a great tool for portraying romance, particularly with reference to moonlight. Some of the better-known titles are 'By the Light of the Silvery Moon' (Edward Madden and Gus Edwards, 1909), 'Moon River' (Andy Williams, music by Henry Mancini and lyrics by Johnny Mercer), 'Blue Moon' (Rodgers and Hart, 1934), 'Fly Me to the Moon' (by Bart Howard, 1954, and often associated with Frank Sinatra). Pink Floyd's *The Dark Side of the Moon* (1973), is one of the greatest selling albums in the Western world. Its title, however, does not refer to the far side of the Moon but to madness and the negative aspects of mental illness. Pink Floyd were familiar with these issues as ex-member Syd Barrett was suffering from serious psychiatric problems at the time.

Classical music is similarly represented. A well-known example is by Ludwig van Beethoven (1770–1827), whose Piano Sonata no. 14 (1801) posthumously became known as the 'Moonlight Sonata' following a remark by the German poet and music critic Ludwig Rellstab (1799–1860), who wrote that the sonata reminded him of the moonlight reflected off Lake Lucerne. Another is by Claude Debussy, who named the third movement of his *Suite bergamasque* (1890–1905) 'Clair de lune' (By the light of the Moon).

Poster for *Moon over Miami* (1941), starring Betty Grable.

The domain of the Queen
of the Night, in an early
19th-century stage design
for Mozart's *The Magic Flute*.

The poetic Moon

Two well-known nursery rhymes concern the Moon. 'Jack and Jill' derives from a Norse myth in which a brother and sister are taken from the Earth to the Moon, where they remain and can be seen carrying a pail, a parallel to the story of the man in the Moon. A more obvious example is the rhyme:

> Hey diddle diddle, the cat and the fiddle,
> The cow jumped over the Moon.
> The little dog laughed to see such fun
> And the dish ran away with the spoon!

The origins of the notion of being 'over the Moon', or in a euphoric state, are lost in the mists of time, but author J.R.R.

A cinematic view of love under the silver Moon, in an early 20th century movie publicity shot.

An illustration, *c.* 1900, for 'Jack and Jill', a nursery rhyme that alludes to the lunar mythology.

Tolkien provides an alternative background to this rhyme in *The Fellowship of the Ring*.[11] The original four-line rhyme is expanded into thirteen five-line stanzas, 'The Man in the Moon Came Down Too Soon', sung by the hobbit Frodo. The man in the Moon gets so drunk that as dawn approaches the landlord, cat, dog, dish and spoon have to roll him up the hill and bundle him back into the Moon. With the Moon so positioned the cow is able to jump over it. The song ends when the Moon rolls behind a hill just before the Sun raises her head.

Another famous rhyme comes from Edward Lear (1812–1888) in 'The Owl and the Pussy Cat':

And hand in hand, on the edge of the sand,
They danced by the light of the Moon,
The Moon,
The Moon,
They danced by the light of the Moon.

In poetry the Moon is a very popular subject, contributing to many nocturnal views and moods. In late antiquity, the Greek poet Sappho (610–570 BC), who is thought to be one of the earliest female poets, wrote about it:

The stars about the lovely Moon,
Fade back and vanish soon,
When, round and full, her silver face,
Swims into sight, and lights all space.'[12]

There have been some very prolific Chinese poets, an example being Tu Fu (712–770), a poet of the Tang dynasty who left around 1,500 poems to posterity. Several of these feature the Moon, including 'Full Moon' which begins, 'Isolate and full, the Moon, / Floats over the house by the River.'[13]

In Europe, apart from a few early examples such as Chaucer, lunar-themed poetry blossomed in the sixteenth century, especially

Edward Lear's 'The Owl and the Pussycat' verses culminate in the marriage of the pair, whereupon, 'hand in hand, on the edge of the sand, / They danced by the light of the Moon'.

Nicholas Hilliard,
Elizabeth I (1586).
This portrait miniature
presents the monarch
as Cynthia, the moon
goddess: there is
a crescent moon
jewel in her hair and
a scattering of arrows
in her ruff, a further
allusion to Diana.

during the reign of Elizabeth I, as the cult of Cynthia the Virgin Queen was at its peak. Edmund Spenser (1552–1599) and Sir Philip Sydney (1554–1586) wrote such poetry, and Ben Jonson (1572–1637), in 'Hymn to Diana', gave us 'Queen and Huntress, chaste and fair, / Now the Sun is laid to sleep, / Seated in thy silver chair'. Sir Philip Sydney, in 'Astrophel and Stella' (1582), wrote, 'With how sad steps, O Moone, thou climb'st the skies! / How silently, and with how wanne a face!'

In Jacobean England, John Milton (1608–1674) in part IV of the epic poem *Paradise Lost* describes dawn thus: 'Nor grateful ev'ning mild; nor silent night, / With this her solemn Bird, nor walk by Moon, / Or glittering star-light, without thee is sweet.' Then Robert Burns (1759–1796), in the 1783 poem 'The Rigs o' Barley', describes a romantic moonlight tryst: 'Beneath the Moon's unclouded night, / I held away to Annie.'

By the eighteenth and nineteenth centuries poetry had become more introspective and sophisticated, with poets such as Samuel Taylor Coleridge (1772–1834) writing several Moon-themed poems, including 'To the Autumnal Moon': 'I watch thy gliding, while with watery light'. Percy Bysshe Shelley (1792–1822) gives us a more jaded view in 'The Waning Moon': 'The Moon arose up in the murky East, / a white and shapeless mass . . . Art thou pale for weariness' – and a romantic view in 'Love's Philosophy':

See! the mountains kiss high heaven,
And the waves clasp one another;
No sister flower would be forgiven,
If it disdained its brother;
And the sunlight clasps the Earth,
And the moonbeams kiss the sea; –
What are all these kissings worth,
If thou kiss not me?

William McGonagall (1825–1902) provides a return to the more populist view with 'The Moon', in which all nine verses start with 'Beautiful Moon, with thy silvery light' and go on to describe the positive aspects of moonlight to cheer the weary traveller, and the lovers in the night.[14] Thomas Hardy (1840–1928), who is better known for his fiction than for his poetry, witnessed a lunar eclipse which inspired the poem, 'At a Lunar Eclipse': 'Thy shadow, Earth, from Pole to Central Sea, / Now steals along upon the Moon's meek shine / In even monochrome and curving line, / Of imperturbable serenity.'[15] Hardy makes further use of this lunar observation as a metaphor for change in his novel *The Return of the Native* (1878). The central characters Eustacia and Clym meet under an eclipsing Moon, and Clym proposes marriage to Eustacia. D. H. Lawrence (1885–1930) wrote several lunar poems, including 'Moonrise': 'And who has seen the Moon, who has not seen / Her rise from out of the chamber of the deep'. During the same period, the classicist A. E. Housman (1859–1936) wrote in *A Shropshire Lad*, 'White in the Moon the long road lies, / The Moon stands blank above; / White in the Moon the long road lies / That leads me from my love'.

From the twentieth century there is e. e. cummings (1894–1962): 'who knows if the Moon's a balloon, coming out of a keen city in the sky – filled with pretty people'. Robert Graves (1895–1985), the popular novelist and poet, wrote in 'The Cruel Moon': 'Nurse says the Moon can drive you mad? / No, that's a silly story, lad! / Though she be angry, though she would / Destroy all England if she could'. W. H. Auden (1907–1973) wrote several lunar poems, one, 'Moon Landing', concerning the 1969 Apollo landings: 'Homer's heroes were certainly no braver, than our trio, but more fortunate: Hector was excused the insult of having his valour covered by television.' The Welsh poet Dylan Thomas (1914–1953) gave us 'In my craft or sullen art / Exercised in the still night / When only the Moon rages / And the lovers lie abed / With all their griefs in their arms / I labour by singing light'.[16] Yorkshire-born Ted Hughes (1930–1998), in his own distinctive style, in 'The Harvest Moon' wrote 'The flame-red Moon, the harvest Moon, / Rolls along the hills, gently bouncing, / A vast balloon, / Till it takes off, and sinks upward / To lie in the bottom of the sky, like a gold doubloon.'

Thus the Moon is firmly embedded in all forms of popular culture and has been used in the creative arts throughout history. Sometimes its physical form and influence on the natural environment have been a feature, but more popularly it has been used as a metaphor for the human condition. Through science fiction it has inspired us to look outwards from Earth and think about extraterrestrial life, stimulating space travel and culminating in Neil Armstrong walking on the Moon in 1969. An adventure spanning many decades, a story of humanity's genius, the journey to the Moon is described in the next chapter.

6 One Small Step for a Man, One Giant Leap for Mankind

After the Second World War and the development of the rockets by the Germans, travel to the Moon became plausible. The rocket opened up space exploration, which started in earnest with the launch of Sputnik 1 by the Soviets in 1957. The space race began between the Cold War superpowers, the USA and USSR (and the UK for a while). The drive to explore the Moon and be the first to land a man on it was to push political and economic developments to new heights for three decades – from the 1950s to the 1970s. For the Americans, the manned Apollo space programme had to be a success as it represented the Western political view of freedom, fought so hard for during the Second World War. Apollo also pushed forward the frontiers of technology, giving us the largest rockets ever made, as well as advancing the frontiers of human endurance and pioneering spirit. The Moon landings had a huge cultural impact around the world and forever altered our view of humanity's place in the universe.

Inspired by the circumnavigation of the globe in the sixteenth century, scientists considered the Moon as the next destination to be explored, and described various designs for 'flying ships'. John Wilkins (1614–1672) published his ideas for a spaceship capable of travel to the Moon in his book *The Discovery of a New World in the Moone* (1638). In these pre-Newtonian days, before the term 'gravity' had been coined, space travel seemed a simple next step.

The era of space travel really started with Konstantin Eduardovich Tsiolkovsky (1857–1935), a Russian visionary and

An attempt to reach the Moon in a balloon by the Chevalier Humguffier and the Marquis de Gull, *c.* 1780.

mathematician who first worked out the trajectory and speed a spacecraft would need to follow to leave the Earth and its atmosphere. He also speculated on the energy content of the fuel needed to produce the necessary amounts of thrust.[1] This was before the rocket as we know it today had been invented. His contribution to rocketry has been recognized by naming a far side lunar crater after him.

During the Second World War the Nazis developed rocket technology with a view to attacking cities in the USA. Under Wernher von Braun's lead they built the first ballistic missile, the progenitor for all modern rockets, called the V2 (Vergeltungswaffe 2: retaliation weapon 2). By the time the war ended around 3,000 V2 rockets had been launched by German forces, and mainly targeted at London and later Antwerp. All V2 rockets were painted black and white, a colour scheme kept by all rocket developers to this day.

By July 1946, in the United States, von Braun and his team, using modified V2s shipped from Germany, had entered space by reaching an altitude of 134 km. Fifteen months later, in October 1947, the Soviet Russians, under the direction of Sergei Korolev, launched their own modified V2s, reaching a higher altitude of 300 km.[2]

By 1949, the maximum altitude reached was 393 km, and both the Americans and the Soviets realized that a single rocket could never provide the lift needed to launch a space capsule big enough to carry humans safely into Earth orbit. Thus began the race to build bigger and more powerful multistage rockets. The Soviets overcame this using a low-tech approach. Having developed their Raketa 7, or R7, rocket, they bolted together several rocket motors to make a more powerful one, giving them a strong early lead. The Americans decided to design new rockets from scratch. These early, more technologically advanced, American rockets either exploded on the launch pad or failed just after launch. The Soviets' lead allowed them to lift the first satellite, Sputnik, into Earth's orbit in October 1957 (a launch date chosen to celebrate the Communist Revolution). Sputnik was tracked by Britain's newly built Jodrell Bank radio telescope in Cheshire. The satellite emitted a weak radio signal that could be picked up on Earth as it passed overhead. Although small and simple, it was the first man-made object to orbit the Earth in space, and the Soviets secured a huge publicity coup. This event marked the point in human history where the space race became a matter of national pride, and our view of the solar system and the universe was changed for ever.

One of the more bizarre American responses to Sputnik is contained in a USAF report from 1958 – the suggestion of detonating a nuclear bomb on the Moon. The motivation was threefold – scientific, military and political. It was planned to send a modified intercontinental ballistic missile to the Moon with a nuclear device in its nose. Upon reaching the Moon it would be detonated and the nuclear flash would be seen from the Earth. The Americans thought this would prove to the rest of the world that they were the world leaders in space technology. At the same time it would provide nuclear scientists with information on how nuclear weapons behaved in space, as both the Soviets and the Americans were looking towards militarizing space. Luckily the project was never taken seriously and peaceful space exploration programmes were chosen instead. Small nuclear devices were later taken to the Moon by both nations to power the Apollo surface experiments and the Soviet robot rovers.

In the following month the Soviets launched Sputnik II, this time with a dog, Laika, on board, who died shortly after launch. The Americans finally achieved space flight with Explorer One in January 1958. At this time NASA (National Aeronautics and Space Administration agency) was born, as the Americans had come to realize that only a single civilian space agency could deliver the coordinated effort to develop manned space flight. Before this the inter-force rivalry between the Army, Navy and Air Force had spread the resources. Within days of the birth of NASA, Project Mercury, with its aim to put a man into space, was created. The Soviets, leading the space race, now focused on the Moon. Luna 1 was launched in January 1959 and passed within 6,000 km of the Moon before ending up orbiting the

Sun. This was followed by Luna 2 in September 1959, which crashed into the lunar surface (Palus Putredinis, or Marsh of Decay), becoming the first man-made object to reach the Moon. With each new launch the level of sophistication increased. Luna 3 (October 1959) orbited the Moon, returning the first images of the far side. Meanwhile the USA launched a craft that passed within 60,000 km of the Moon.[3]

On 12 April 1961 the USSR gained another first: Yuri Gagarin became the first human in space. In Vostok 1, he completed a single orbit in one hour and 48 minutes before returning to Earth. The Americans were still behind, struggling with their smaller but more sophisticated designs, which led to more launch failures and lower weight limits. The Soviets, with technically inferior but much larger rockets, were not so constrained. On their return to Earth the Soviet cosmonauts would eject from their capsule at 4,000 m and land by parachute, usually on Soviet territory.

On 25 May 1961 President John F. Kennedy, in an attempt to leapfrog the Soviets' lead and following Fidel Castro's May Day speech to Congress, famously announced the USA's commitment to place a man on the Moon:

> I believe that this nation should commit itself to achieving the goal, before this decade is out, of landing a man on the Moon and returning him safely to Earth. No single space project in this period will be more impressive to mankind, or more important for the long range exploration of space: and none will be so difficult or expensive to accomplish.

For the USA progress was slow. Ranger 1 and 2 (1961) remained in Earth orbit; Ranger 3 (1962) passed within 32,000 km of the Moon. Later in the year Ranger 4 crashed on the far side of the Moon, while in October 1962 Ranger 5 travelled uneventfully past the Moon. It was Ranger 6 that gave the USA its first partial success by landing on the Moon but failing to return any pictures.[4] Rangers 7, 8 and 9 returned thousands of close-up pictures of the Moon's surface, showing that it was suitable for a small spacecraft to land on.

The space race began to quicken in pace when in May 1961, 23 days after the Soviets, and live on television, Alan Shepard became the first American in space, but not orbit. This launch was quickly followed in 1961 by Gherman Titov in Vostok 2, who spent 25 hours in space orbiting the Earth more than seventeen times, while in 1962 John Glenn became the first American to orbit the Earth, three times in total.

The USA was trailing in the space race and in 1962 President John F. Kennedy, while visiting Rice University, gave a now famous speech, emphasizing the importance of the USA going to the Moon:

> We choose to go to the Moon in this decade and do the other things – not because they are easy, but because they are hard, because that goal will serve to organize and measure the best of our energies and skills, because that challenge is one that we are willing to accept, one we are unwilling to postpone, and one which we intend to win.

On 16 June 1963 the USSR claimed another first by launching the first woman into space, 26-year-old Valantina Tereshkova. In Vostok 6 she spent around three days in orbit.[5] In 1963 President Kennedy was assassinated and the Mercury programme ended.

Between 1964 and 1965 the Soviets made even greater advances into space, but the USA was catching up fast, building ever better, bigger and more powerful rockets. With mission control now in Houston, the new Gemini programme began to deliver American successes. In June 1965, the first U.S. spacewalk – by Edward White – was achieved. Then later, Gemini V, crewed by L. Gordon Cooper Jr and Charles Conrad Jr, broke the space endurance record by staying in orbit for nearly eight days. This was followed in December by the endurance record being extended to two weeks, with Gemini VII completing over 200 orbits, and eleven days into the mission Gemini VI rendezvoused with Gemini VII. The mission was a great success with the two capsules, although not touching each other, spending around five hours alongside each other.

overleaf:
An artist's impression of the lunar landscape at sunset, in Agnes Giberne's *Sun, Moon and Stars* (1884).

While the Americans were testing the life support systems needed to get astronauts to the Moon and back, the Soviets were concentrating on robotic missions and in February 1966 launched Luna 9, successfully landing a 100 kg craft onto the Ocean of Storms (Procellum Maria). After seven hours the craft began to beam the first ever close-up TV images of the lunar surface back to Earth. The transmissions were intercepted by British astronomers at the Jodrell Bank radio telescope, who realized that the signals were in the same format as those used by newspapers to transmit images electronically by fax. The observatory informed the *Daily Express,* which was able to publish the first images of the lunar surface worldwide, barely before the Soviets had seen them themselves.[6] Whether the use of this format by the Soviets was a deliberate ploy has never been established. One thing is for certain, however – these images, circulated worldwide, ended centuries of speculation, showing the Moon's surface to be barren, covered in a fine dust and, more importantly, without plants or any sign of life or water. The landing also confirmed that the surface was robust enough to withstand the weight of a spacecraft.

Four months after Luna 9, the Americans launched the Surveyor craft, which also landed on the Ocean of Storms. This craft had higher-resolution colour cameras, but as the Moon is largely grey this did not add much to what Luna 9 had observed. Further craft were launched: Surveyor 2 failed, while Surveyor 3 landed and, equipped with a remotely controlled digging arm, was able to dig about 1 cm into the loose dusty surface before hitting the solid regolith beneath. This craft sent 6,400 images back to Earth before it ran out of power, remaining untouched until the Apollo 12 crew landed nearby and brought its camera back to Earth.[7] In 1967, Surveyor 4 crashed into the Moon, while Surveyor 5, launched in September, landed in the Sea of Tranquillity from where it sent thousands of pictures and performed some geological analysis, confirming that the rocks were ancient basaltic lava. The last two Surveyor craft, 6 and 7, sent back 118,000 high-resolution images. Some have suggested that these must be the dullest and most expensive images ever collected.

Through the Gemini programme the Americans had developed the space technology necessary for a manned Moon landing. The final stages of this journey began with the Apollo programme. The programme was named after Greek god of the Sun, who every day drove his chariot across the sky to bring light to the world. The NASA manager Abe Silverstein gave his own reason for choosing Apollo: 'Apollo riding across the Sun was appropriate to the grand scale of the proposed program.'[8] Given that the target was the Moon, rather than the Sun, Diana or Cynthia, as goddess of the Moon, might have been a more appropriate choice of name. A new rocket, the most powerful and immense one ever built, was at the core of the Apollo programme. The Saturn rocket consisted of three stages. The first stage was capable of generating a massive 3.5 million kg of thrust and of lifting three astronauts into orbit along with 36 tons of payload or spacecraft. To do this the rocket motors had to burn a total of 2,000 tons of fuel at a rate of thirteen tons per second, accelerating the rocket from zero to nine times the speed of sound (around 10,000 km per hour) in under three minutes.[9]

On 27 January 1967, three astronauts, Virgil Grissom, Edward White and Roger Chaffee, were preparing to launch in Apollo 1 when a fire began in the capsule, which was full of pressurized oxygen. Within minutes an inferno had consumed the inside of the capsule, killing all three. They were the first fatalities of the space race and American ambitions were delayed as a result. This setback for the USA spurred on President Brezhnev, who now pushed the Soviet space programme which had been falling behind. On 23 April 1967 the cosmonaut Vladimir Komarov was launched into orbit but before the mission started the craft began experiencing technical difficulties and he decided to abandon the mission. Unable to establish the correct re-entry trajectory and use his parachute, he subsequently plummeted to the ground at 650 km per hour, becoming the USSR's first fatality. Meanwhile, Apollo 4, 5 and 6 were unmanned missions used to test the launch systems and re-entry protocols. Apollo 7 was the next Saturn rocket to carry astronauts. The Soviets returned to using animals as cosmonauts and

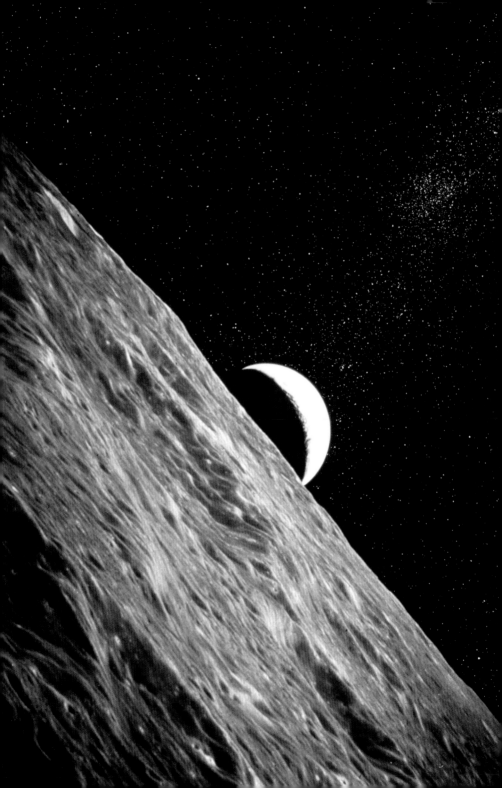

in September 1968 sent Zond 5 around the Moon. The craft contained the first terrestrial creatures to visit the Moon – two turtles and some flies and worms. They suffered no observable ill health from the flight.

Eventually Apollo 8 launched on 21 December 1968 and the three astronauts Frank Borman, James Lovell and William Anders became the first humans to enter true space. Their mission was to orbit the Moon. To great publicity they entered lunar orbit late on Christmas Eve. By Christmas Day they had completed ten orbits, giving the American public the first televised view of the Moon's surface as they woke up on Christmas morning. The far side was not very newsworthy, but a more memorable sight was the green, brown, blue and white Earth rising above the monochrome horizon of the Moon, all set against the black backdrop of space, and described by James Lovell as 'a grand oasis in the big vastness of space'.[10] These images had a huge cultural impact, making people the world over aware of the Earth's unique place in the universe. This view spawned many new environmental groups and movements, promoting the message of the Earth being humankind's only home, and one that we must all share and value. This first image of Earthrise has probably been seen by virtually everybody in the world. The success of the Apollo 8 provided a positive end to a year in which the Vietnam War was still raging, a year marked by the Soviet invasion of Prague, and the assassinations of Martin Luther King in April and Robert F. Kennedy in June.

The Soviets still had hopes of reaching the Moon first, and in January 1969 they scored another first by successfully transferring cosmonauts between Soyuz 4 and 5 while in orbit. The Soviets had been slowly developing their own rocket, which was potentially mightier than the Saturn v. Their rocket, the N-1, was 100 m high, with four stages. It had 30 motors in the first stage, compared to Saturn's five. The first unmanned launch in February 1969 failed after a minute, while a second launch exploded on the launch pad, destroying a third N-1 rocket stored nearby. This was catastrophic for the USSR, ending their chances of a manned Moon landing.

Earth rising, as seen beyond the lunar surface, photographed on the Apollo Mission of 1969.

The Apollo 9 and 10 missions launched in March and May 1969 respectively were test missions, rehearsing transfer between the lunar module and command module in Earth orbit. In July Apollo 11 was ready on the launch pad, waiting to take three astronauts, Neil Armstrong, Michael Collins and Edwin (Buzz) Aldrin into lunar orbit. The launch on 16 July and the early part of the mission went to plan, and after four days they entered lunar orbit, disappearing behind the Moon on Monday 19 July. The next day Neil Armstrong and Buzz Aldrin left the command module, *Columbia*, and entered the lunar module, *Eagle*. After a long series of tests the Moon landing began. For the *Eagle* to land it needed to decelerate as it descended to 2,000 m above the surface from an orbital speed of 6,000 km per hour to just 100 km per hour. Finally the *Eagle* had to slow to a halt so that it could land by hovering gently and dropping onto the surface. The flight was to be automatically controlled using the primitive on-board computer as the fuel supply was limited. Everything went to plan until, at an altitude of 610 m, an alarm sounded. Aldrin, Armstrong and mission control at Houston all had no idea what the warning was for. At 305 m the error light was still on but mission control had finally found out what the fault code was (simply put, it was a signal that the landing computer was overloaded) and gave the go-ahead to land. The *Eagle* was heading for a crater with lots of big boulders strewn across its surface and to prevent disaster Neil Armstrong took over manual control, deciding to land 300 m further on. Although this would add only a few seconds to the flight time it could dangerously limit the fuel supply. At an altitude of 76 m there was only 94 seconds' worth of fuel left. They soon passed the 15 m point, a point of no return, as below this point there was insufficient time to turn off the descent engine and switch on the ascent engine to return the *Eagle* back into orbit. Neither was the *Eagle* strong enough to drop onto the surface from this height so for the astronauts to survive Armstrong would have to land quickly and carefully. As they hovered at 9 m, they looked out of the window for a landing site, but they had a limited view of the surface which was further obscured as the rocket

Earthrise taken by the Apollo 11 astronauts as they emerged from the far side of the Moon.

thrust spat up dust and debris. They were landing in the morning, so the angle of the Sun was low, reinforcing any shadows. At this point they had only 30 seconds' worth of fuel left. The two astronauts were looking for a blue light on their dashboard which would illuminate on touchdown. With only 10 seconds' worth of fuel left, the blue light finally came on, and Armstrong switched the engine off. He had made a perfect landing. Suddenly the tension was over, and complete silence descended over the astronauts, the only sound coming from mission control over the intercom: 'We copy you down, *Eagle.*' Neil Armstrong, no longer following protocol, announced: 'Engine arm is off . . . Houston, Tranquillity Base here. The *Eagle* has landed.' Many years later Neil Armstrong said this was the emotional high point of the mission. The two astronauts said nothing and just slapped each other on the shoulders.

After three hours the astronauts were ready for their moonwalk and Armstrong in his spacesuit mounted the ladder on the outside of the lunar lander, where an externally mounted camera allowed everyone on Earth to see his movements live. The picture was extremely poor, and provided a somewhat ghostly image of the astronauts. Eventually Armstrong was on the last rung of the ladder and was given permission to step onto the surface. As he stepped down, he said, 'That's one small step for [a] man, one giant leap for mankind.' Amazingly it was left up to Armstrong to think up this now famous line for this momentous point in human history. He described the surface as firm and covered in a fine and powdery dust that stuck to his boots. The idea that he might sink into the lunar surface was at the forefront of his mind.

Fifteen minutes later it was Buzz Aldrin's turn to leave the *Eagle*, which he did while being filmed by Armstrong. There are no photographs of Armstrong himself on the lunar surface. There was only one camera and Armstrong took all the pictures. Once on the surface, their suits functioned perfectly, and the astronauts soon learnt how to move around. They were given the task of collecting rock and soil samples, erecting the 'Stars and Stripes' (which the *Eagle* blew over as it took off) and placing an instrument package that was to be left behind. This contained a seismometer and a laser reflector which would allow terrestrial astronomers to measure the distance between the Earth and Moon to the nearest centimetre. After 2 hours and 47 minutes on the surface the astronauts returned to the *Eagle*. The samples were stored in a sealed box since no one was sure of their toxicity levels. They then had time for some sleep and altogether were on the surface for 21 hours. As they went through prelaunch checks they found that the ascent engine switch had been broken off, probably as they struggled into their space suits before their Moonwalk. If this switch was inoperable then the *Eagle* was going nowhere. However, this potential disaster was averted when someone at Houston control realized that the hollow ends of their ballpoint pens would fit over the remaining metal pin. Thus, on 21 July 1969, the *Eagle* lifted safely off the lunar surface.

In the Nixon Libraries of the u.s. National Archives can be found a speech written for President Nixon in the event that the first Moon landing failed:

> Fate has ordained that the men who went to the Moon
> to explore in peace will stay on the Moon to rest in peace.
> These brave men, Neil Armstrong and Edwin Aldrin, know
> that there is no hope for their recovery. But they also know
> that there is hope for mankind in their sacrifice . . . there is
> some corner of another world that is forever mankind.

Fortunately this speech was never required.

After completing the rendezvous with *Columbia*, the *Eagle* was jettisoned and the three-man crew slept while their craft sped towards Earth. They reached Earth on the morning of 24 July. Ahead of them was re-entry, the most dangerous part of the mission. If they did not hit the atmosphere at the correct angle then *Columbia* would break up and if the heat shield failed the passengers would be roasted alive. Everything went to plan and at 11.51 a.m. the capsule splashed down in the Pacific Ocean. The whole mission had lasted just over eight days, during which they had travelled 1.5 million km and returned with 21.5 kg of Moon rocks and soil. Most importantly of all they had left Earth and walked on extraterrestrial land. To many around the world this was an event that changed civilization for ever, signalling a new beginning and leaving an indelible mark on our history. The disappointed Soviets did not even cover the live Moon landing and only mentioned it after it had happened.

Four months later, on 14 November 1969, Apollo 12 was launched. This mission lasted ten days and the two astronauts

overleaf:
A close-up of the lunar surface, an expanse of dust and large rocks photographed in 1971.

(Charles Conrad and Alan Bean) landed in the Oceanus Procellum (Ocean of Storms), where they completed two Moonwalks, spending around seven hours altogether outside the lunar module, *Intrepid*. They landed next to the Surveyor 3 craft which had been there for 31 months. As well as 34 kg of geological samples, they also brought back bits of the Surveyor for examination, including the TV. This time the astronauts described running on the Moon as being like a giraffe running in slow motion.[11] Again a set of instruments was left behind and the *Intrepid*, once the astronauts had finished with it, was deliberately crashed into the Moon so that the seismometers could measure the Moon's response.

On 11 April 1970, Apollo 13 launched at thirteen minutes past the hour, with astronauts James Lovell, Jack Swigert and Fred Haise on board. Late on 13 April, as the crew went through their final checks before going to sleep, disaster struck. One of the two oxygen storage tanks ruptured. A few weeks before the launch two wires had been damaged, losing their insulation, which now created a spark arcing between them, causing the oxygen to explode. The explosion was silent as it occurred in the vacuum of space, although a shudder was felt inside the craft. On realizing what had happened, Swigert radioed Houston with the now famous remark 'Houston, we've had a problem.' The oxygen was used to drive the fuel cells that powered the two craft. After the explosion, only one of the three fuel cells was working, so the mission to the Moon had to be abandoned. The extent of the damage was not appreciated at first, but as checks were done it soon became apparent that the astronauts might not have enough oxygen to power the fuel cells and keep them alive. The only way home was to loop around the Moon back to Earth, a journey of six days, an event dramatized in the film *Apollo 13*. To save power for landing, the crew had to sit in the cold and dark, conserving water and oxygen. As they passed behind the Moon they were 400,000 km away from the Earth, the furthest man has ever travelled from this planet. The astronauts reached home safely on 17 April.

In November of that year the Soviets launched the Luna 17 mission and successfully landed a 754 kg robot lander, named

Lunokhod, in the Sea of Rains. It spent ten months on the Moon and travelled over 10 km. While adding little to lunar knowledge. it was the progenitor of all future planetary missions. Curiosity is one of the latest robot landers and is exploring Mars.

On 30 January 1971 Apollo 14 launched, tasked with landing in the Fra Mauro Highlands, an area littered with debris from the ancient impact that created the Imbrium Basin. The two astronauts, Alan Shepard (the first American in space and the first to play lunar golf) and Edgar Mitchell, completed two Moonwalks, this time covered by colour TV images, lasting around nine hours in total. They returned safely on 9 February, carrying 43 kg of geological samples. Now with three lunar landings under their belt, the Americans were confident that they had overcome all the technical difficulties. The Soviets still dreamed of landing on the Moon but in June 1971 they lost three cosmonauts in Earth orbit. The USSR had stretched their technology to its limits and had to concede that a manned flight to the Moon was beyond their capabilities.

Apollo 15 was launched 26 July 1971, and now missions were driven by scientific objectives, with the astronaut's role changing from test pilot to lunar explorer. The landing site chosen was the Mare Imbrium (Sea of Rains), next to the Apennine Mountain Range. This time the lunar module carried a battery-powered lunar rover and landed in Hadley Rille, a 400-m-deep valley carved out by an ancient magma flow. After some sleep, an excursion was taken in the lunar rover to St George Crater 10 km away, followed in the next two days by a second and third excursion, including a trip up a lunar mountainside, totalling 28 km. This region was chosen as it was thought to contain the oldest (billions of years) lunar rocks. Using drills and the rover, the astronauts were able to collect a range of interesting samples, including a green glassy rock that had been formed during an ancient meteor impact and rocks that were 4.15 billion years old. They also found a huge cinder cone, providing the first direct evidence of explosive lunar volcanism. On their departure, the astronauts David Scott and James Irwin left a commemorative plaque in memory of the fourteen NASA and USSR cosmonauts who had died during the development of each of the nations' space vehicles. The mission ended safely on 7 August. This was considered one of the most revealing missions, returning to Earth with 80 kg of lunar samples.

Continuing in the scientific vein, Apollo 16 was launched nine months later, in April 1972. A mountainous region, the Descartes Highlands, was chosen as the landing site, which the mission reached on 21 April. During their three-day stay the astronauts spent around 21 hours on the lunar surface and completed three excursions in the lunar rover, a total of 27 km. This time the mission returned around 100 kg of geological samples. The material was not of volcanic origin as predicted but consisted of impact breccia, a mixture of granules solidified into a solid rock looking a bit like concrete. The seismometer, with great fortuity, picked up a meteor impact as it slammed into the far side of the Moon near Mare Moscoviense.[12]

The last lunar mission was Apollo 17. Although twenty missions had been planned, the last three were cancelled because of the cost, but also because there was little political gain to be derived from further lunar exploration. These included the cancellation of a rocket-powered Mooncopter that would have been used to traverse the lunar surface. Apollo 17 left the launch pad on 7 December 1972, landing four days later at the Taurus-Littrow valley, a site expected to provide some interesting geological samples both from old rock and from recently impacted rocks. Here the soil sparkled, as it was full of glass particles formed during the impact of a meteor. Later orange and red soil was found, tinted by microscopic glass beads containing high concentrations of titanium. In total the mission spent 74 hours on the surface and covered 35 km, collecting 110 kg of material. Aware that his was the last scheduled mission, on leaving the Moon, astronaut Eugene Cernan said, 'As we leave the Moon at Taurus-Littrow, we leave as we came and, God willing, as we shall return, with peace and hope for all mankind.' Mission control read a statement provided by President Nixon: 'As Challenger leaves the surface of the Moon, we are conscious not of what we leave behind, but of what lies before us.'[13] The

The last Apollo
mission, 17. Harrison
Schmitt is seen with
the lunar rover.

last and longest Apollo mission ended safely, splashing down on 19 December 1972. This ended both the Apollo programme and the space race.

Over a three-year period, twelve men had been to the Moon at a cost of $25 billion. All returned safely to Earth. Today only eight are still alive and are in their late seventies or older. In total they spent around 300 hours on the lunar surface and brought back nearly half a ton of geological samples. The robotic USSR missions returned one-third of a kilogram of material.

The Apollo missions give us the phrase, 'If they can put a man on the Moon, why can't they . . .?' While the technological feats of engineering and the scientific aspects of the Moon landings were explained to the public, this failed to capture their imag-inations. Lunar geology was a subject that did not enrapture the public, and soon interest waned. The TV networks, despite improved colour images, dropped live coverage of the later Apollo missions. By then the goal of landing the first men on the Moon had been achieved and to all it was now just a dead, dusty planet. All eyes had turned to Mars, where the hope of finding a more terrestrial landscape and even alien life forms beckoned.

After Apollo the Soviets could see that there was little political gain in landing a cosmonaut on the Moon, and they still lacked a sufficiently powerful booster rocket to lift large pay-loads into lunar orbit.[14] Instead they sought to leapfrog the Americans by developing robotic rovers and then perhaps land-ing men on Mars. While U.S. astronauts were visiting the Moon the USSR explored it remotely. In September 1971 Luna 18 reached the Moon as directed, but just as it was about to land in the Sea of Fertility it ran out of fuel and crash landed on the lunar surface.[15] Some scientific information was gleaned from the mission – the density of the lunar surface was determined by looking at the impact. Luna 19 followed immediately. This was an orbiting craft which, rather than landing, was designed to image the surface from orbit in order to identify future landing sites and measure and map the inhomogeneous magnetic fields or mascons (areas of anomalous magnetism). The craft had com-pleted 4,000 orbits by October 1972. In February 1972 Luna 20 landed in Apollonius, an upland area north of the Sea of Fertility. The on-board television camera sent pictures back of the surface while a drill got to work and collected a 30 g sample of lunar rock, which was sent back to Earth in its own spacecraft for analysis. In January 1973 Luna 21 reached the Moon. Now that the Apollo programme had ceased, the USSR had the Moon to itself. Landing in the remains of the eroded Monnier crater, on the edge of the Sea of Serenity, the battery-powered rover Lunokhod 2, weighing close to 850 kg and their most sophisticated robotic rover yet, set out across the lunar surface. During its six-month operation it travelled around 37 km. The mission ended early on 3 June when the heat exchangers became clogged with lunar soil and the rover overheated. To have operated over such a long period, when temperatures ranged over 300°c, between night and day, was at the time a triumph of engineering.

The vehicle was doing many of the things that the manned missions did, and demonstrated that robotic explorers were a much safer and cheaper alternative. The following year, in June 1973, Luna 22 launched and orbited the Moon until November 1975. This set the scene for the next set of landing missions. In 1974, Luna 23 was badly damaged when it crash landed in the Sea of Crisis. The next mission, Luna 24, landed in the same area in August 1976. The mission returned 170 g of material. Luna 25 was to carry Lunokhod 3 to the Moon, but the mission was cancelled because of the lack of sufficiently powerful launcher – and funds – marking the end of the Soviet Union's interest in the Moon. Their sights were now set on Venus.

There is much still to do on the Moon, such as visiting the poles and the far side. No human has left Earth's orbit for the past 40 years. Why have we not gone back to the Moon? The USSR lost interest as there were no political gains to be had, especially once it was decided that setting up a lunar colony was far beyond Soviet budget or technology. The Americans gave up once the USSR had lost interest. No one then believed that Apollo 17 would be the last manned lunar mission of the twentieth century. Mars and Venus were considered the next targets, and the USA began sending unmanned probes to Mars in preparation.

An astrolabe of 1577, incorporating the lunar cycle.

Although the space race was the civilian face of the Cold War and associated arms race, its end did not mark the end of the Cold War, which came in 1989 with the breaching of the Berlin Wall. The rocket technology developed during the space race allowed both sides to develop missiles capable of carrying multiple nuclear warheads across continents. The simultaneous deployment of these so-called ICBMs (intercontinental ballistic missiles) would ensure global Armageddon or MAD (mutually assured destruction). Fortunately, with the ending of the Cold War, thousands of these weapons have been decommissioned. Today the developing nations, such as North Korea, give space exploration as a reason for developing rockets.

Until now the Moon has only been considered a goal reachable with state funding, but with the space shuttle cancelled, the USA is trying to commercialize space, encouraging industry and entrepreneurs to develop the next generation of rockets and space vehicles. The costs are high but the potential benefits huge, on a par with the opening up of North America in the seventeenth and eighteenth centuries. History has shown that once an initial discovery has been made, interest wanes as ideas of immediate benefit fail to be realized. After time a new generation of explorers see new benefits and exploration recommences. With the emergent Asian economies and commercial exploitation encouraged by the usa, a new lunar age is dawning.

7 Legacy

Within our solar system there are hundreds of moons – Saturn alone has over 60 – and every year new ones are discovered. With the discovery of hundreds of new planets orbiting distant stars, so called exo-planets, it is only a matter of time before we discover new exo-moons. The Moon has influenced our vocabulary and is used as a synonym for mysterious, pale and milky objects. A good example is moonstone, a mineral and gemstone with a white translucent sheen that was thought by the ancient Romans to be made of solidified moonbeams. Both the Romans and Greeks associated the gem with their gods. Selenium, a rare Earth element (symbol Se) discovered in 1817, was named after the Moon. Its sister element, tellurium (symbol Te), had been named after the Earth in the 1780s. As fate would have it, selenium deserves this name as it is sensitive to ambient light such as moonlight, with its electrical resistance changing with light intensity in a phenomenon known as photoresistivity. The visibility of the Moon is also dependent on its exposure to sunlight. Selenium is an essential micronutrient element necessary to maintain human health. Unlike selenium, the Moon is no longer considered to affect our health, and its influence is relegated to fantasy and the occult, such as witchcraft. Our cultural view of the Moon is shifting in that it is no longer an object of mystery but one ripe for exploration.

The space race ended with the Apollo missions and when the Soviets turned their gaze towards Venus and Mars. The Moon was abandoned as a dry, lifeless orb – it had failed to grab

A Victorian moon brooch of 1888, with a crescent moon and shooting-star frame consisting of diamonds with a carved moonstone face.

the public's attention and politicians saw no merit to be gained from funding further missions. Then, as the twentieth century drew to a close, a new generation of scientists began to consider the Moon worthy of interest and two NASA lunar orbital missions were launched: Clementine in 1994 and the Lunar Prospector in 1998. This interest has increased in the twenty-first century, with a renaissance of Moon exploration. It began with the European Space Agency (ESA), who finally funded their first mission to the Moon, SMART 1, in 2003. This spacecraft was designed to be more of a test vehicle than a lunar explorer, and ended its days being deliberately crashed into the Lacus Excellentiae (Lake of Excellence) on the far side of the Moon. Asian economies such as Japan, China and India had observed the rapid technological advances engendered by the USA and USSR space programmes and have developed their own rockets. Such rockets can fulfil two potential goals – reaching space

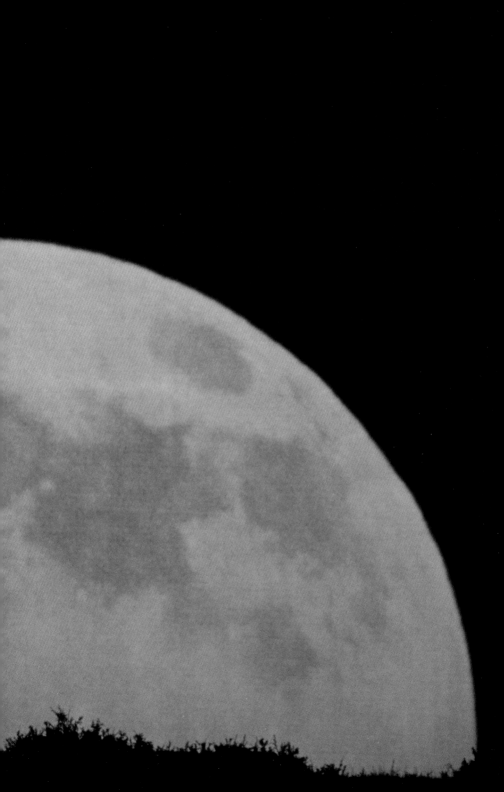

and delivering a nuclear warhead to anywhere in the world. Having a national space programme fulfils national pride and casts political leaders in a positive light. Japan sent the Hiten satellite around the Moon in January 1990 and then the more sophisticated Kaguya (SELENE) in 2007. One month later China launched Chang'e I into lunar orbit, sent to test the technology more than any specific exploration. Chang'e II was launched in 2010, this time serving as a test for a more ambitious unmanned lunar landing mission – Chang'e III, which landed on the Moon on 14 December 2013 at the Mare Imbrium (Sea of Rains). Accompanying the lander was a small, 140 kg lunar rover named Yutu, or 'Jade Rabbit'. Over two days the four-wheeled rover travelled around 50 m before it broke down. The Chinese had become the third nation to land a craft on the Moon, and plan a manned Moon landing for 2025. The Chang'e programme is named after a Chinese legend in which a young goddess flies to the Moon.

India's space programme began with the launch of Chandra-yaan 1 (Sanskrit for Moon craft) in 2008, which successfully orbited the Moon for two years and returned thousands of high-resolution images of the lunar surface. India is also planning a manned lunar mission with help from the Russians, who have more powerful rockets.

The USA has not given up on the Moon. NASA have launched four missions recently – the Lunar Reconnaissance Orbiter and LCROSS missions in 2009, followed in 2011 by the Gravity Recovery and Interior Laboratory (GRAIL) mission with its twin spacecraft which orbited the Moon together, used to determine the mass and structure of the lunar interior. Like its predecessors, the two probes met their end by being crashed into the surface. The Lunar Atmosphere and Dust Environment Explorer (LADEE) mission was launched on 6 September 2013 and reached its planned lunar orbit on 20 November 2013. It was a successful mission that collected new information about the Moon's exosphere and its dust composition.

While these science-based missions are backed by NASA, there is little political will to return to the Moon, although

Presidents Bush and Obama both made positive statements about reviving a manned Moon landing programme – albeit only during election time.

The interest has shifted away from government-sponsored missions to private funding by wealthy benefactors. Following in the footsteps of the Orteig challenge in 1919, when a $25,000 prize was offered to the first person to fly non-stop from New York to Paris, and won by the outsider Charles Lindbergh in the *Spirit of St Louis*, the Google Lunar x prize was offered in the same spirit in 2007. The $30 million prize will go to the first privately funded team to land a craft on the Moon by 2015. To be eligible for the prize the craft must travel 500 m on the lunar surface and send high-definition video back to Earth. Originally 34 teams from across the world registered and took up the challenge, but as time passes several teams have dropped out and still no one has claimed the prize. Many of these private funders are visionaries wishing to commercialize the Moon, for instance with the creation of lunar hotels. With a gravity of one-sixth of that of the Earth, tourists would experience a strength-to-body-weight ratio six times greater than on Earth. The first gymnasts on the Moon would be able to perform feats only dreamed of on Earth, such as a quadruple twist, for example.

Others wish to exploit the Moon's vast mineral wealth of substances such as helium3 and titanium, or to use it as a staging post for longer manned missions to Mars and beyond. One of the more bizarre plans involves moving asteroids from the asteroid belt into a close lunar orbit from where their mineral wealth can be mined and transported to Earth. Only time will tell whether we are witnessing a new era in the Moon's history and a new beginning in humankind's extraterrestrial future.

This silvery orb, as itself, as Selene or as the man in the Moon, continues to fascinate us, its waxing and waning light providing a magical sheen to our culture, not only lighting the night skies but lighting humankind's future, and our expansion and colonization of the solar system and the stars beyond.

TIMELINE FOR MOON LANDINGS

AD 160   Lucian of Samosata writes the first story, called *True Story*, about a fictional journey to the Moon

1634   John Wilkins publishes his ideas for a spaceship capable of travel to the Moon in his book *The Discovery of a New World in the Moone*

1865   Jules Verne in his work of fiction *From the Earth to the Moon* describes the great speed or escape velocity needed to overcome the Earth's gravitational pull. A giant cannon is used to shoot a shell-shaped spaceship containing three men and a dog on a one-way mission to the Moon

1940s   The V2 single stage rocket developed as weapons of mass destruction by the Germans, who learn how guide their rockets

1959   Luna 1 and Pioneer 4 fly close to the Moon, while Luna 2 crashes onto the surface. Luna 3 photographs the far side of the Moon

1961   John F. Kennedy asks Congress to fund a lunar landing mission by the end of the decade

1965   Ranger 9 sends back images of the lunar surface

1966   Surveyor 1 and Luna 13 land on the Moon and return thousands of detailed images

1967   A fire in Apollo 1 kills astronauts and Surveyor 6 achieves the first rocket take-off from the lunar surface

1968   Apollo 7, the first manned mission to test the command module in space, is launched. The first live TV broadcast from space is made. Apollo 8 is the first manned mission to orbit the Moon, observing the far side and the earthrise over the Moon

1969   Apollo 11 lands on the Moon

1970   Apollo 13: mission aborted, the crew return safely after using the Moon's gravity to swing their craft back to Earth

1971   Apollo 14 explores the Fra Mauro uplands with a lunar
        surface littered with many ancient small impact craters.
        Apollo 15 uses the lunar rover for the first time

1972   Apollo 16, the first mission with a geological focus, lands
        on the Descartes Highlands. Apollo 17, the last manned lunar
        mission to the Taurus-Littrow area, is launched

1973   The USA launches Skylab, the world's first space station

1975   The Apollo-Soyuz programme begins a joint operation
        between the USSR and USA, marking the end of the space race.

1976   The last Russian probe, Luna 24, lands on the Moon and
        returns lunar samples to Earth

1990   The Japanese spacecraft Hiten orbits the Moon

1994   The American Clementine orbiter takes high-definition
        images of the lunar surface

1998   The American Lunar Prospector orbiter, placed in low polar
        orbit, measures lunar gravity and magnetism

2003   The European Space Agency's only lunar mission, the orbiter
        Smart 1, is used as a test-bed for space technology

2007   The Japanese Selene lunar orbiter surveys the surface
        distribution of the mineral Olivine. The Chinese Chang'e 1
        lunar orbiter is launched as test of technological ability.

2008   The Indian Chandrayaan 1 lunar orbiter mission is launched

2009   The USA's NASA LRO lunar orbiter maps day and night
        surface temperatures. The NASA LCROSS finds frozen water
        on the Moon

2010   The Chinese Chang'e II surveys the Moon for possible
        landing sites for future missions

2011   The USA's GRAIL lunar orbiter increases our knowledge of
        the structure of the Moon's interior

2013   The USA's LADEE lunar orbiter explores the lunar exosphere
        and its dust content. The Chinese Chang'e III returns to
        the Moon, successfully landing and deploying a lunar rover

REFERENCES

Introduction

1 Patrick Moore, *Guide to the Moon* (London, 1953), p. 125.
2 Quoted in Carol Ann Duffy, ed., *To the Moon: An Anthology of Lunar Poems* (London, 2009), p. 17.

1 Lunar Shadows

1 Junjun Zhang et al., 'The Proto-Earth as a Significant Source of Lunar Material', *Nature Geoscience*, v/4 (March 2012), pp. 251–5.
2 Andreas Reufer et al, 'A Hit-and-run Giant Impact Scenario', *Icarus*, ccxxi/1 (July 2012), pp. 296–9.
3 Oded Aharonson et al., 'Why Do We See the Man in the Moon?', *Icarus*, ccxix/1 (March 2012), pp. 241–3.
4 Katherine H. Joy et al., *Science*, 'Direct Detection of Projectile Relics from the End of the Lunar Basin-forming Epoch', 336/6087 (June 2012), pp. 1426–9.
5 R. A. Bamford et al., 'Mini-magnetospheres above the Lunar Surface and the Formation of Lunar Swirls', *Physical Review Letters*, cix/8 (July 2012), pp. 1101–06.
6 Reufer et al., 'Hit-and-run'.
7 J. Andrew McGovern et al., 'Mapping and Characterization of Non-polar Permanent Shadows on the Lunar Surface', *Icarus*, ccxxiii/1 (January 2013), pp. 566–81.
8 Patrick Moore, *Guide to the Moon* (London, 1953), pp. 73–4.
9 Hamish Lindsay, *Tracking Apollo to the Moon* (London, 2001), p. 11.
10 Mark A. Wieczorek et al., 'The Crust of the Moon as seen by GRAIL', *Science*, cccxxxix/6120 (February 2013), pp. 671–5.
11 Dag Linnarsson et al., 'Toxicity of Lunar Dust', *Planetary and Space Science*, lxxiv/1 (December 2012), pp. 57–71.
12 Timothy Harley, *Moon Lore* (London, 1885), pp. 227–58.
13 Moore, *Guide to the Moon*, p. 44.

14 Lindsay, *Tracking Apollo*, p. 20.
15 James Hamilton, *Volcano* (London, 2012), p. 16.
16 Moore, *Guide to the Moon*, p. 115.
17 William Herschel and Joseph Banks, 'An Account of Three Volcanoes in the Moon', *Philosophical Transactions of the Royal Society of London*, 77 (1787), pp. 229–32.
18 Christian Koeberl, 'Craters on the Moon from Galileo to Wegener: A Short History of the Impact Hypothesis, and Implications for the Study of Terrestrial Impact Craters', *Earth, Moon and Planets*, 85–6 (2001), pp. 209–24.
19 Christian Koeberl, 'The Late Heavy Bombardment in the Inner Solar System: Is there any Connection to Kuiper Belt Objects?', *Earth, Moon and Planets*, 92 (2004), pp. 79–87.
20 Aharonson et al., 'Man in the Moon', pp. 241–3.

2 Time and Motion

1 R. G. Foster and T. Roenneberg, 'Human Responses to the Geophysical Daily, Annual and Lunar Cycles', *Current Biology*, xviii/7 (September 2008), pp. 784–94.
2 Stewart Ross, *Moon: Apollo 11 and Beyond . . . The Ultimate Guide to our Nearest Neighbour* (Oxford, 2009), p. 114.
3 David Ewing Duncan, *The Calendar: The 5000-year Struggle to Align the Clock and the Heavens, and What Happened to the Missing Ten Days* (London, 1998), pp. 65–72.
4 Lisa Jardine, *Ingenious Pursuits: Building the Scientific Revolution* (London, 1999), pp. 178–82.
5 Patrick Moore, *Guide to the Moon* (London, 1953), p. 143.
6 Ibid., p. 144.
7 Colin A. Ronan, *The Cambridge Illustrated History of the World's Science* (Cambridge, 1983), pp. 118–20.
8 Tristan Gooley, *The Natural Navigator* (London, 2010), pp. 138–52.
9 Jaroslaw Wlodarczyk, 'Libration of the Moon: Hevelius's Theory, and its Early Reception In England', *Journal for the History of Astronomy*, 149 (November 2011), pp. 495–519.
10 Moore, *Guide to the Moon*, pp. 135–40.
11 Oded Aharonson, Peter Goldreich and Re'em Sari, 'Why Do We See the Man in the Moon?', *Icarus*, ccxix/1 (March 2012), pp. 241–3.
12 Alina Iosif and Bruce Ballon, 'Over and Above: Bad Moon Rising: The Persistent Belief in Lunar Connections to Madness', *Canadian Medical Association Journal*, clxxiii/12 (January 2005), pp. 1498–500.

13 M. F. Sterzik, S. Bagnulo and E. Palle, 'Biosignatures as Revealed by Spectropolarimetry of Earthshine', *Nature*, CDLXXXIII/7387 (March 2012), pp. 64–6.

14 Michael Carlowicz, *The Moon* (New York, 2007), p. 86.

15 Quoted in Carol Ann Duffy, ed., *To the Moon: An Anthology of Lunar Poems* (London, 2009), p. 64.

16 Russell Doescher, Don Olson and Roger Sinnot, 'Did the Moon sink the *Titanic?*', *Sky and Telescope*, 123 (April 2012), pp. 34–9.

17 Ian Stewart, *Seventeen Equations that Changed the World* (London, 2012), p. 291.

18 Gavin Pretor-Pinney, *The Wavewatcher's Companion* (London, 2011), p. 256.

19 Moore, *Guide to the Moon*, pp. 31–2.

20 Ibid.

21 Gooley, *The Natural Navigator*, pp. 138–52.

22 J. T. Desaguliers, 'An Attempt to Explain the Phaenomenon of the Horizontal Moon Appearing Bigger, Than When Elevated Many Degrees above the Horizon: Supported by an Experiment, Communicated Jan. 30, 1734–5', *Philosophical Transactions*, 39 (1735), pp. 436–44; Samuel Dunn, 'An Attempt to Assign the Cause: Why the Sun and Moon Appear to the Naked Eye Larger When They Are Near the Horizon. With an Account of Several Natural Phenomena', *Philosophical Transactions*, 52 (1761), pp. 462–73.

23 Joseph Antonides and Toshiro Kubota, 'Binocular Disparity as an Explanation for the Moon Illusion', arXiv.org (January 2013), 1301.2715.

3 Stories and Legends

1 'Endymion', *Encyclopaedia Britannica*, 11th edn (Cambridge, 1911), vol. IX, p. 390.

2 Timothy Harley, *Moon Lore* (London, 1885), p. 49; Tom Holland, *In the Shadow of the Sword: The Battle for Global Empire and the End of an Ancient World* (London, 2012), p. 108.

3 The King James Bible, Isaiah 49:12.

4 Ken Dowden, *European Paganism: The Realities of Cult from Antiquity to the Middle Ages* (London, 2000).

5 Ibid.

6 Harley, *Moon Lore*, p. 39.

7 Ibid., p. 95.

8 Sarah Gristwood, *Arbella: England's Lost Queen* (London, 2003), p. 102.

9 Harley, *Moon Lore*, p. 63.

10 Niall McCrae, *The Moon and Madness* (Exeter, 2011), p. 30.

11 Sabine Baring-Gould, *The Book of Were-Wolves: Being an Account of Terrible Superstition* [1865] (London, 1973), p. 267.

12 Will Summers and R. J Harris, *R. J. Harris's Moon Gardening*, 2nd edn (Shrewsbury, 2007), p. 256.

13 'On the Supposed Influence of the Moon on the State of the Weather', *The Chronicle*, vol. 1 (1836), pp. 60–75.

14 Trea Martyn, *Elizabeth in the Garden: A Story of Love, Rivalry and Spectacular Design* (London, 2008), pp. 159–61.

15 Ibid., p. 242.

16 Roy Strong, *Gloriana: The Portraits of Queen Elizabeth I* (London, 2003), pp. 125–8.

17 Harley, *Moon Lore*, p. 7.

18 Ibid., p. 9.

19 Ibid., p. 14.

20 Ibid., p. 31.

21 Ibid., p. 37.

22 Miriam R. Levy, 'How the Man got in the Moon', *Pedagogical Seminary*, 3 (1895), pp. 317–18.

23 James Hamilton, *Volcano* (London, 2012), p. 83.

24 Michael Carlowicz, *The Moon* (New York, 2007), p. 59.

25 Ibid., p. 58.

26 Basavaprabhu Achappa et al., 'Rituals Can Kill: A Fatal Case of Brucine Poisoning', *Australian Medical Journal*, v/8 (September 2012), pp. 421–3.

4 Man and the Moon

1 Oren Froy, 'Circadian Rhythms, Aging, and Life Span in Mammals', *Physiology*, 26 (August 2011), pp. 225–35.

2 S. P. Law, 'The Regulation of the Menstrual Cycle and its Relationship to the Moon', *Acta Obstetricia et Gynecologica Scandinavica*, LXV/1 (1986), pp. 45–8; I. Ilias et al., 'Do Lunar Phases Influence Menstruation? A Year-long Retrospective Study', *Endocrine Regulations*, XLVII/3 (July 2013), pp. 121–2.

3 Niall McCrae, *The Moon and Madness* (Exeter, Devon, 2011), pp. 28–30.

4 Ibid., p. 35.

5 Ibid., pp. 39–40.

6 Ibid., p. 40.

7 Sarah Gristwood, *Elizabeth and Leicester* (London, 2008), p. 409.

8 Anna Whitelock, *Elizabeth's Bedfellows: An Intimate History of the Queen's Court* (London, 2013), pp. 28–9.

9 Thomas Wright, *Circulation: William Harvey's Revolutionary Idea* (London, 2012), p. 190.

10 Alina Iosif and Bruce Ballon, 'Over and Above: Bad Moon Rising:
    The Persistent Belief in Lunar Connections to Madness',
    Canadian Medical Association Journal, CLXXIII/12 (January 2005),
    pp. 1498–500.

11 Madeleine M. Grigg-Damberger and Nancy Foldvary-Schaefer,
    'Diagnostic Yield of Sleep and Sleep Deprivation on the EEG in
    Epilepsy', *Sleep Medicine Clinics*, VII/1 (January 2012), pp. 91–8.

12 Iosif and Ballon, 'Over and Above'.

13 McCrae, *The Moon and Madness*, p. 102.

14 James Rotton and I. W. Kelly, 'Much Ado about the Full Moon:
    A Meta-Analysis of Lunar-lunacy Research', *Psychological Bulletin*,
    XCVII/2 (March 1985), pp. 286–306.

15 Martin Voracek et al., 'Not Carried Away by a Moonlight Shadow:
    No Evidence for Associations between Suicide Occurrence and
    Lunar Phase among more than 65,000 Suicide Cases in Austria
    1970–2006', *Wiener Klinische Wochenschrift*, CXX/11–12 (June 2008),
    pp. 343–9.

16 F. Amaddeo et al., 'Frequency of Contact with Community-based
    Psychiatric Services and the Lunar Cycle: A 10-year Case-Register
    Study', *Social Psychiatry and Psychiatric Epidemiology*, XXXII/6
    (August 1997), pp. 323–6.

17 Geneviève Belleville et al., 'Impact of Seasonal and Lunar Cycles
    on Psychological Symptoms in the ED: An Empirical Investigation
    of Widely Spread Beliefs', *General Hospital Psychiatry*, XXXV/2
    (October 2012), pp. 192–4.

18 M Röösli et al., 'Sleepless Night, the Moon is Bright: Longitudinal
    Study of Lunar Phase and Sleep', *Journal of Sleep Research*, XV/2
    (June 2006), pp. 149–53.

19 S. Baxendale and J. Fisher, 'Moonstruck? The Effect of the Lunar
    Cycle on Seizures', *Epilepsy and Behaviour*, XIII/3 (October 2008),
    pp. 549–50.

20 C. L. Raison, H. M. Klein and M. Steckler, 'The Moon and
    Madness Reconsidered', *Journal of Affective Disorders*, LIII/1
    (May 1999), pp. 99–106; R. G. Foster and T. Roenneberg, 'Human
    Responses to the Geophysical Daily, Annual and Lunar Cycles',
    Current Biology, XVIII/17 (September 2008), pp. 784–94; Christian
    Cajochen et al., 'Evidence that the Lunar Cycle Influences Human
    Sleep', *Current Biology*, XXIII/15 (July 2013), pp. 1–4.

21 R. G. Holzheimer, C. Nitz and U. Gresser, 'Lunar Phase does not
    Influence Surgical Quality', *European Journal of Medical Research*,
    VIII/9 (October 2003), pp. 414–18.

22 C. Peters-Engl et al., 'Lunar Phases and Survival of Breast Cancer
    Patients – A Statistical Analysis of 3,757 Cases', *Breast Cancer
    Research and Treatment*, LXX/2 (November 2001), pp. 131–5.

23 A. Kuehnl et al., 'The Dark Side of the Moon: Impact of Moon
    Phases on Long-term Survival, Mortality and Morbidity of
    Surgery for Lung Cancer', *European Journal of Medical Research*, 14
    (April 2009), pp. 178–81; M. May, K. P. Braun, C. Helke, W.
    Richter, H. Vogler, B. Hoschke and M. Siegsmund, 'Lunar Phases
    and Zodiac Signs do not Influence Quality of Radical Cystectomy
    – A Statistical Analysis of 452 Patients with Invasive Bladder
    Cancer', *International Urology and Nephrology*, xxxix/4 (March
    2007), pp. 1023–33.

24 Jeffery H. Shuhaiber et al., 'The Influence of Seasons and Lunar
    Cycle on Hospital Outcomes Following Ascending Aortic
    Dissection Repair', *Interactive Cardiovascular and Thoracic Surgery*,
    10 (2013) pp. 1–5; Rajan Kanth, Richard L. Berg and Shereif
    H. Rezkalla, 'Impact of Lunar Phase on the Incidence of Cardiac
    Events', *World Journal of Cardiovascular Disease*, ii/3 (July 2012),
    pp. 124–8.

25 A. K. Exadaktylos et al., 'The Moon and Stones: Can the Moon's
    Attractive Forces Cause Renal Colic?', *Journal of Emergency
    Medicine*, xxii/3 (April 2002), pp. 303–05.

26 Hojjat Molaee et al., 'The Lunar Cycle: Effects of Full Moon
    on Renal Colic', *Urology Journal*, viii/2 (April 2011), pp. 137–40;
    Sanjay Kalra, Tushar Bandgar and Manisha Sahay, 'The Sun, the
    Moon and Renal Endocrinology', *Indian Journal of Endocrinology
    and Metabolism*, xvi/2 (March 2012), pp. 156–7.

27 Jill M. Arliss et al., 'The Effect of the Lunar Cycle on Frequency
    of Births and Birth Complications', *American Journal of Obstetrics
    and Gynecology*, cxcii/5 (May 2005), pp. 1462–4.

28 Ismini Staboulidou et al., 'The Influence of Lunar Cycle on
    Frequency of Birth, Birth Complications, Neonatal Outcome
    and the Gender: A Retrospective Analysis', *Acta Obstetricia et
    Gynecologica Scandinavica*, lxxxvii/8 (August 2008), pp. 875–9.

29 S. Das et al., 'Do Lunar Phases Affect Conception Rates in
    Assisted Reproduction?', *Journal of Assisted Reproduction and
    Genetics*, xxii/1 (January 2005), pp. 15–18.

30 Keren Agay-Shay et al., 'Periodicity and Time Trends in the
    Prevalence of Total Births and Conceptions with Congenital
    Malformations among Jews and Muslims in Israel, 1999–2006:
    A Time Series Study of 823,966 Births', *Birth Defects Research*
    (Part A), xciv/6 (June 2012), pp. 438–48.

31 Angela Megumi Ochiai et al., 'Atmospheric Conditions, Lunar
    Phases, and Childbirth: A Multivariate Analysis', *International
    Journal of Biometeorology*, lvi/4 (July 2011), pp. 661–7.

32 J. W. Ballantyne, 'The Term "Mooncalf"', *British Medical Journal*
    (1900), pp. 780–81.

5 Lunar Art and Literature

1 Timothy Harley, *Moon Lore* (London, 1885), p. 20.
2 Bernd Brunner, *Moon: A Brief History* (New Haven, CT, and London, 2010), pp. 109–10.
3 Patrick Moore, *Guide to the Moon* (London, 1953), pp. 160–61.
4 Ibid.
5 Hamish Lindsay, *Tracking Apollo to the Moon* (London, 2001), p. 4.
6 David A. Kirby, 'Science Consultants, Fictional Films, and Scientific Practice', *Social Studies of Science*, XXXIII/2 (April 2003), pp. 231–68.
7 Ibid.
8 Dona A. Jalufka and Christian Koeberl, 'Moonstruck: How Realistic is the Moon Depicted in Classic Science Fiction Films?', *Earth, Moon and Planets*, 85–6 (2001), pp. 179–200.
9 Ibid.
10 Trea Martyn, *Elizabeth in the Garden: A Story of Love, Rivalry and Spectacular Design* (London, 2008), p. 89.
11 J.R.R. Tolkien, *The Fellowship of the Ring* (London, 1954), chapter Nine.
12 Carol Ann Duffy, ed., *To the Moon: An Anthology of Lunar Poems* (London, 2009), p. 1.
13 Ibid., p. 27.
14 Ibid., p. 60.
15 Ibid., p. 65.
16 Ibid., p. 136.

6 One Small Step for a Man, One Giant Leap for Mankind

1 John B. West, 'Historical Aspects of the Early Soviet/Russian Manned Space Program', *Journal of Applied Physiology*, XCI/4 (October 2001), pp. 1501–11.
2 James Harford, *Korolev: How One Man Masterminded the Soviet Drive to Beat America to the Moon* (New York, 1997), pp. 64–90.
3 Hamish Lindsay, *Tracking Apollo to the Moon* (London, 2001), p. 20.
4 Ibid., p. 24.
5 Ibid., p. 87.
6 Ibid., p. 119.
7 Ibid., p. 137.
8 Lewis Research Center, NASA, News Release 69-36, Cleveland, July 1960.
9 Lindsay, *Tracking Apollo to the Moon*, p. 215.
10 Alan Shepard and Deke Slayton, *Moon Shot: The Inside Story of America's Apollo Moon Landings* (New York, 2011).

11 Lindsay, *Tracking Apollo to the Moon*, p. 328.
12 David West Reynolds, *Apollo: The Epic Journey to the Moon,
    1963–1972* (London, 2013), pp. 222–61.
13 Lindsay, *Tracking Apollo to the Moon*, p. 20.
14 Donald Beattie, *Taking Science to the Moon: Lunar Experiments and
    the Apollo Program* (Baltimore, MD, and London, 2001), pp. 272–3.
15 Brian Harvey, *Soviet and Russian Lunar Exploration* (Berlin, 2007),
    p. 262.

SELECT BIBLIOGRAPHY

Beattie, Donald, *Taking Science to the Moon: Lunar Experiments and the Apollo Program* (Baltimore and London, 2001)

Brunner, Bernd, *Moon: A Brief History* (New Haven, CT, and London, 2010)

Carlowicz, Michael, *The Moon* (New York, 2007)

Duffy, Carol Ann, ed., *To the Moon: An Anthology of Lunar Poems* (London, 2009)

Elger, Thomas Gwyn, *The Moon: A Full Description and Map of its Principal Physical Features* (London, 1895)

Grego, Peter, *Moon Observer's Guide* (London, 2010)

Harford, James, *Korolev: How One Man Masterminded the Soviet Drive to Beat America to the Moon* (New York, 1997)

Harley, Timothy, *Moon Lore* (London, 1885)

Harvey, Brian, *Soviet and Russian Lunar Exploration* (Berlin, 2007)

Hergé, *The Adventures of Tintin: Destination Moon*, trans. Leslie Lonsdale-Cooper and Michael Turner (London, 2011)

Hergé, *The Adventures of Tintin: Explorers on the Moon*, trans. Leslie Lonsdale-Cooper and Michael Turner (London, 2011)

Lindsay, Hamish, *Tracking Apollo to the Moon* (London, 2001)

McCrae, Niall, *The Moon and Madness* (Exeter, 2011)

Moore, Patrick, *Guide to the Moon* (London, 1953)

Raymond, John, *Men on the Moon: NASA's Project Apollo* (London, 1964)

Ross, Stewart, *Moon: Apollo 11 and Beyond . . . The Ultimate Guide to our Nearest Neighbour* (Oxford, 2009)

Verne, Jules, *From the Earth to the Moon/Around the Moon* (Ware, 2011)

Wells, H. G., *The First Men in the Moon* (London, 2011)

Whitaker, Ewen, *Mapping and Naming the Moon: A History of Lunar Cartography and Nomenclature* (Cambridge, 2003)

Wilkins, John, *The Discovery of a World in the Moone* (London, 1638)

ASSOCIATIONS AND WEBSITES

Societies

American Astronomical Society
www.aas.org

The Astronomical Society of Australia
www.asa.astronomy.org.au

British Astronomical Association Lunar Section
www.baalunarsection.org.uk

European Space Agency
www.esa.int/ESA

Indian Space Research Organisation
www.isro.org

Royal Astronomical Society of New Zealand
www.rasnz.org.nz

Other Moon Information

For a map of the Moon go to the Apollo Image Archive, Arizona State University
www.wms.lroc.asu.edu/Apollo

Moon Connection, full of lunar facts and links to lunar products
www.moonconnection.com

The site of the National Aeronautics and Space Administration (NASA) contains lots of facts and images on the Moon and its exploration
http://nssdc.gsfc.nasa.gov/planetary/planets/Moonpage

Cheryl Robertson's Whole Moon page is full of links to all things to do
with the Moon
www.moonlightsys.com/themoon

Society for Popular Astronomy, Lunar Section, UK
www.popastro.com

ACKNOWLEDGEMENTS

Writing this book was a great pleasure and reflects my lifelong interest in the Moon, from observing the Moon in the sky to reading about its exploration and cultural history. As far as I am aware all the information contained herein is correct.

I would like to thank the staff at Reaktion Books, particularly Michael Leaman, picture editor Harry Gilonis and the text editors.

I would like to thank Karen and Laura for listening to my many lunar musings. Without their encouragement and understanding this book would not have been written.

PHOTO ACKNOWLEDGEMENTS

The author and publishers wish to express their thanks to the below sources of illustrative material and / or permission to reproduce it:

The American Indian Health Care Association: p. 129 (foot); courtesy of the author: pp. 6, 9, 57, 62, 85; photos British Library, London / Robana / Rex Features: pp. 103, 147; British Museum, London (photos © The Trustees of the British Museum): pp. 15, 16, 42, 70, 123 (foot), 134, 135, 136; photos Canadian Press / Rex Features: pp. 22 (foot), 61; photo Chameleons Eye / Rex Features: p. 87; photo Hermann Dobler / imagebroker / image-broker.net / Super stock: p. 72; photo Encyclopaedia Britannica / Rex Features: p. 74; photos Everett Collection / Rex Features: pp. 26, 55, 78, 117, 120, 138, 140, 166–7; photo Camille Flammarion: p. 49; photo Global Warming Images / Rex Features: p. 130; photo Image Broker / Rex Features: p. 52; photo Nils Jorgensen / Rex Features p. 28 (top); photo KeystoneUSA-ZUMA / Rex Features: p. 36; photo Matti Kolho / Rex Features: p. 39; photo Gerard Lacz / Rex Features: pp. 76–7; photo Michael Leaman / Reaktion Books: p. 65 (foot); Look and Learn Picture Library / Brian Christie Collection: p. 83; Musée d'Orsay, Paris: p. 123 (top); Museum of Modern Art, New York p. 125; photo NASA images: p. 22 (top); image NASA / JPL-Caltech / IPGP: p. 25; photo NASA / JPL / Ted Stryk: p. 14; photo NASA / JSC / Arizona State University: p. 21; photo NASA / Rex Features: pp. 162–3; National Gallery, Parma: p. 69; National Maritime Museum, London: pp. 32, 47, 170; National Museum, Copenhagen: p. 65 (top); National Museum, Delhi: p. 126; National Portrait Gallery, London: p. 81; Philadelphia Museum of Art: p. 124; photo Photoservice Electa / Rex Features: p. 69; photo Photoservice Electa / Universal Images Group / Rex Features: p. 40; photo Ann Pickford / Rex Features: p. 73; photo Didier Prix / Sunset / Rex Features: p. 20; photo Rex Features: pp. 158–9; photos Stock Connection / Rex Features: pp. 66, 174–5; photos Universal History Archive / Rex Features: pp. 110, 141; photos Universal History Archive / Universal Images Group / Rex Features: